Solutions Manual

Inorganic Chemistry

Third Edition

Gary L. Miessler • Donald A. Tarr
Saint Olaf College

PEARSON
Prentice Hall

Upper Saddle River, NJ 07458

Project Manager: Kristen Kaiser
Senior Editor: Nicole Folchetti
Editor-in-Chief, Science: John Challice
Vice President of Production & Manufacturing: David W. Riccardi
Executive Managing Editor: Kathleen Schiaparelli
Assistant Managing Editor: Becca Richter
Production Editor: Elizabeth Klug
Supplement Cover Manager: Paul Gourhan
Supplement Cover Designer: Joanne Alexandris
Manufacturing Buyer: Ilene Kahn

Pearson Prentice Hall
Pearson Education, Inc.
Upper Saddle River, NJ 07458

Printed in the United States of America

10 9 8 7 6 5

ISBN 0-13-111246-5

Pearson Education Ltd., *London*
Pearson Education Australia Pty. Ltd., *Sydney*
Pearson Education Singapore, Pte. Ltd.
Pearson Education North Asia Ltd., *Hong Kong*
Pearson Education Canada, Inc., *Toronto*
Pearson Educación de Mexico, S.A. de C.V.
Pearson Education—Japan, *Tokyo*
Pearson Education Malaysia, Pte. Ltd.

Contents

Preface

This manual was prepared for use either by instructors or by students. The answers given are more detailed than are likely to be needed by instructors, but are not intended to explore every aspect of a problem. In many cases, reference to the original or review literature will show that a problem is much more complex than it seems at first glance, and we recommend such referrals highly.

We personally prefer to provide our students with answers after they have attempted to work out their own solutions. However, we recognize that for some students, the availability of complete answers is a great help in their study, and have written this manual with such use in mind. The danger is that the student may refer too quickly to these answers, and not think about the problem enough; such use reduces the real usefulness of this book and makes independence in thinking about inorganic chemistry more difficult.

We welcome suggestions from readers for improvement, either in the original text or in this manual. Such suggestions make changes in future editions more useful and help remove errors and ambiguities.

Donald A. Tarr
Gary L. Miessler

Most of the calculations referred to in this manual were done with Spartan Pro, Version 6.0.6. A few were done with CACHe.

Chapter 2
Atomic Structure

2-1 **a.** $\lambda = \dfrac{h}{mv} = \dfrac{6.626\times10^{-34}\,\text{J s}}{9.110\times10^{-31}\,\text{kg}\times0.1\times2.998\times10^{8}\,\text{m s}^{-1}}$

$= 2.426\times10^{-11}\,\text{m}$

b. $\lambda = \dfrac{h}{mv} = \dfrac{6.626\times10^{-34}\,\text{J s}}{0.400\,\text{kg}\times0.1\times2.998\times(10\,\text{km/hr}\times10^{3}\,\text{m/km}\times1\,\text{hr}/3600\,\text{s}}$

$= 6\times10^{-34}\,\text{m}$

2-2 $E = R\left(\dfrac{1}{2^2} - \dfrac{1}{n_h^2}\right)$; $R = 1.097\times10^{7}\,\text{m}^{-1} = 1.097\times10^{5}\,\text{cm}^{-1}$; $\lambda = \dfrac{hc}{E} = \dfrac{1}{\bar{\nu}}$

$n_h = 4$; $E = R\left(\dfrac{1}{4} - \dfrac{1}{16}\right) = R\left(\dfrac{12}{64}\right) = 20{,}570\,\text{cm}^{-1} = 4.085\times10^{-19}\,\text{J}$;

$\lambda = 4.862\times10^{-5}\,\text{cm} = 486.2\ \text{nm}$

$n_h = 5$; $E = R\left(\dfrac{1}{4} - \dfrac{1}{25}\right) = R\left(\dfrac{21}{100}\right) = 23{,}040\,\text{cm}^{-1} = 4.577\times10^{-19}\,\text{J}$;

$\lambda = 4.341\times10^{-5}\,\text{cm} = 434.1\ \text{nm}$

$n_h = 6$; $E = R\left(\dfrac{1}{4} - \dfrac{1}{36}\right) = R\left(\dfrac{8}{36}\right) = 24{,}380\ \text{cm}^{-1} = 4.841\times10^{-19}\,\text{J}$;

$\lambda = 4.102\times10^{-5}\,\text{cm} = 410.2\ \text{nm}$

2-3 $E = R\left(\dfrac{1}{4} - \dfrac{1}{49}\right) = R\left(\dfrac{45}{196}\right) = 25{,}190\,\text{cm}^{-1} = 5.002\times10^{-19}\,\text{J}$

$\lambda = \dfrac{1}{\bar{\nu}} = 3.970\times10^{-5}\,\text{cm} = 397.0\ \text{nm}$

2-4 **a.**

$$-\frac{h^2}{8\pi^2 m}\frac{\partial^2\Psi}{\partial x^2} = E\,\Psi$$

$$\Psi = A\sin rx + B\cos sx$$

$$\frac{\partial\Psi}{\partial x} = Ar\cos rx - Bs\sin sx$$

$$\frac{\partial^2\Psi}{\partial x^2} = -Ar^2\sin rx - Bs^2\cos sx + \frac{h^2}{8\pi^2 m}(Ar^2\sin rx + Bs^2\cos sx)$$

$$= E(A\sin rx + B\cos sx)$$

If this is true, then the coefficients of the sine and cosine terms must be independently equal:

$$\frac{h^2Ar^2}{8\pi^2 m} = EA\;;\quad \frac{h^2Bs^2}{8\pi^2 m} = EB$$

$$r^2 = s^2 = \frac{8\pi^2 mE}{h^2}\;;\quad r = s = \sqrt{2mE}\,\frac{2\pi}{h}$$

b.

$$\Psi = A\sin rx;\quad \text{when } x = 0, \Psi = A\sin 0 = 0$$

$$\text{when } x = a, \Psi = A\sin ra = 0$$

$$\therefore ra = \pm n\pi; r = \pm\frac{n\pi}{a}$$

c.

$$E = \frac{r^2h^2}{8\pi^2 m} = \frac{n^2\pi^2}{a^2}\frac{h^2}{8\pi^2 m} = \frac{n^2h^2}{8ma^2}$$

d.

$$\int_0^a \Psi^2 dx = \int_0^a A^2\sin^2\left(\frac{n\pi x}{a}\right)dx = A^2\frac{a}{n\pi}\int_0^a \sin^2\left(\frac{n\pi x}{a}\right)d\left(\frac{n\pi x}{a}\right)$$

$$= \frac{a}{n\pi}A^2\left[\frac{1}{2}\left(\frac{n\pi x}{a}\right) - \frac{1}{4}\sin\left(\frac{2n\pi x}{a}\right)\right]_0^a = 1$$

$$= \frac{aA^2}{n\pi}\left[\frac{1}{2}\frac{n\pi a}{a} - \frac{1}{4}\sin 2n\pi - 0 + \frac{1}{4}\sin 0\right] = 1$$

$$A = \sqrt{\frac{2}{a}}$$

2-5 **a.** $3p_z$

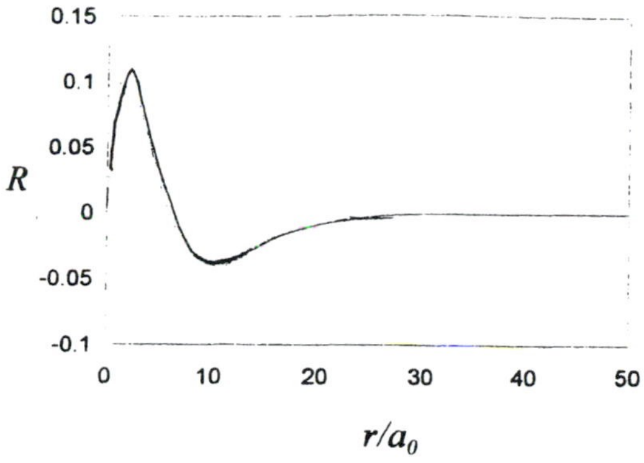

a. $4d_{xz}$

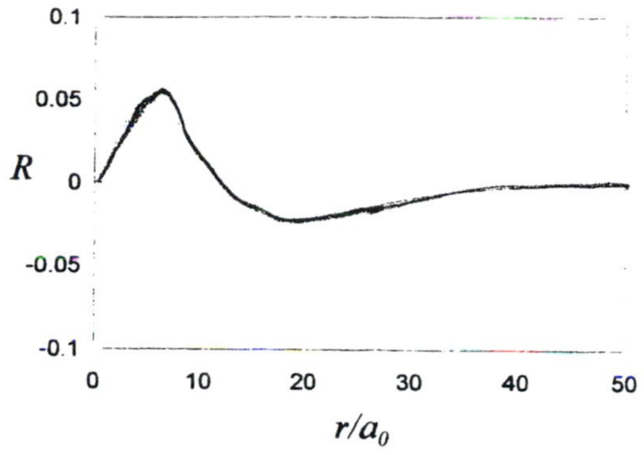

b. $3p_z$

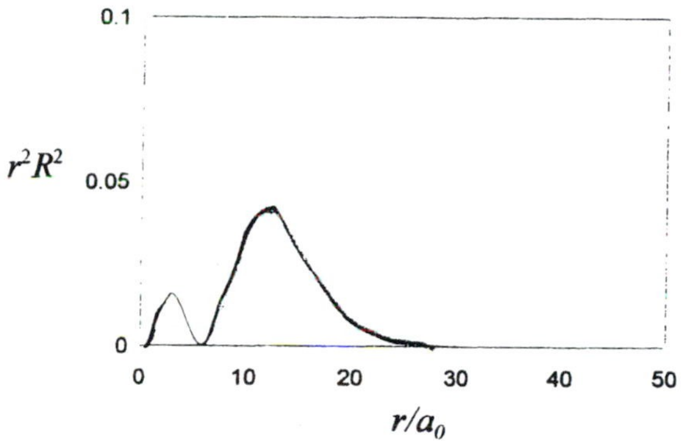

b. $4d_{xz}$

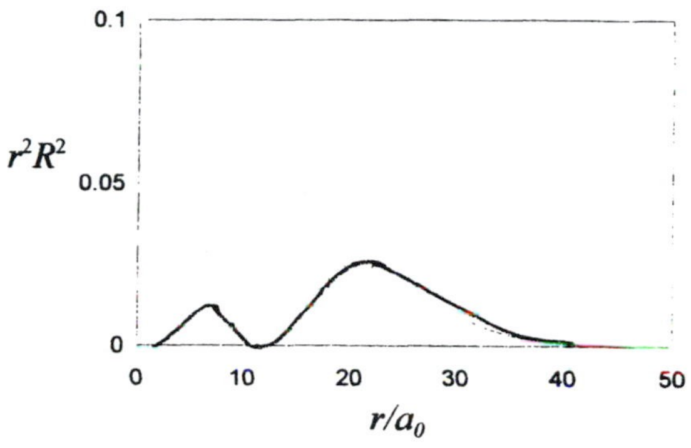

c. $3p_z$

Contours
(dashed lines
are nodes)

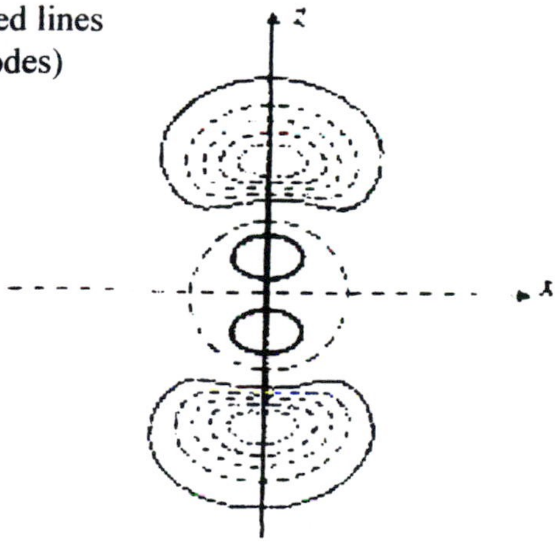

c. $4d_{xz}$

Contours
(dashed lines
are nodes)

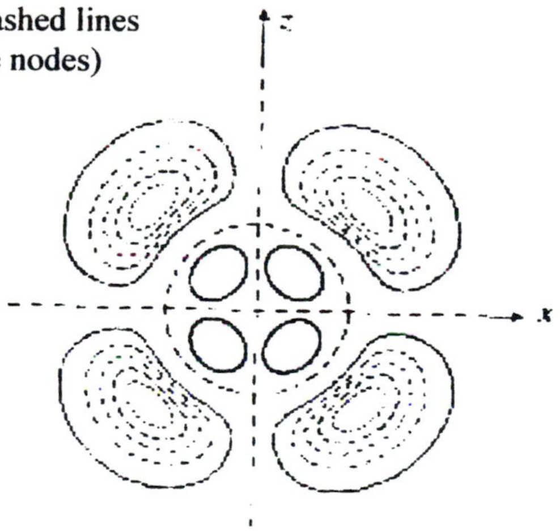

2-6 **a.** $4s$

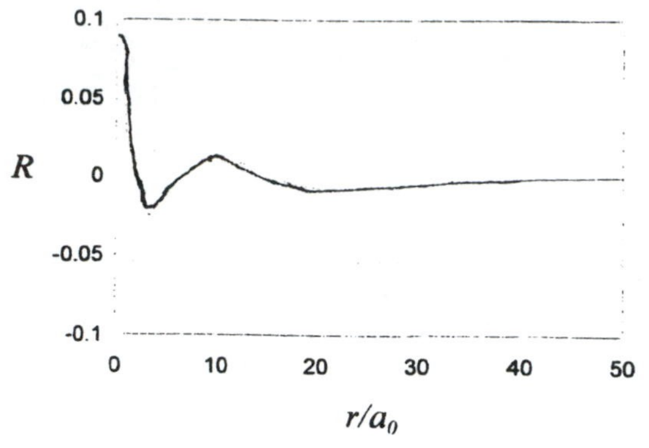

a. $5d_{x^2-y^2}$

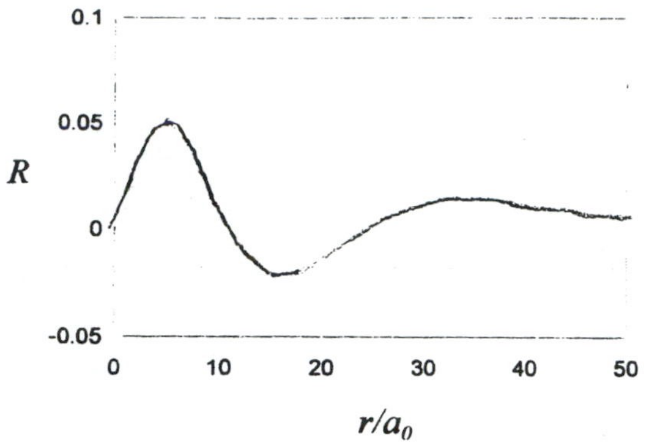

b. $4s$

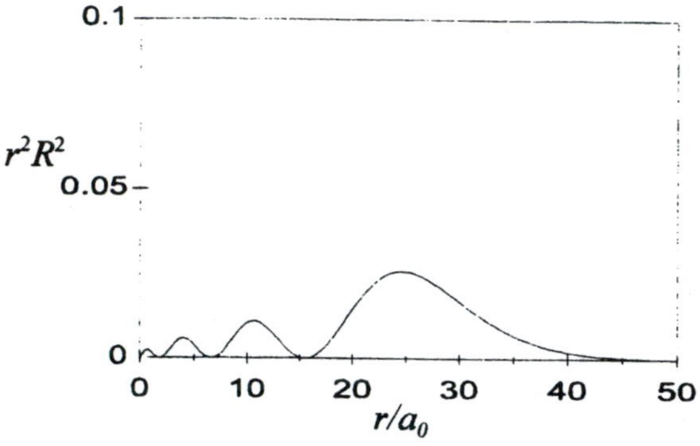

b. $5d_{x^2-y^2}$

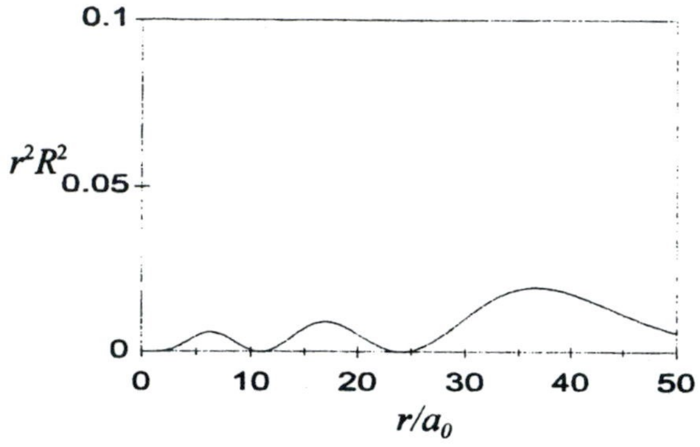

c. $4s$

Contours (dashed lines are nodes)

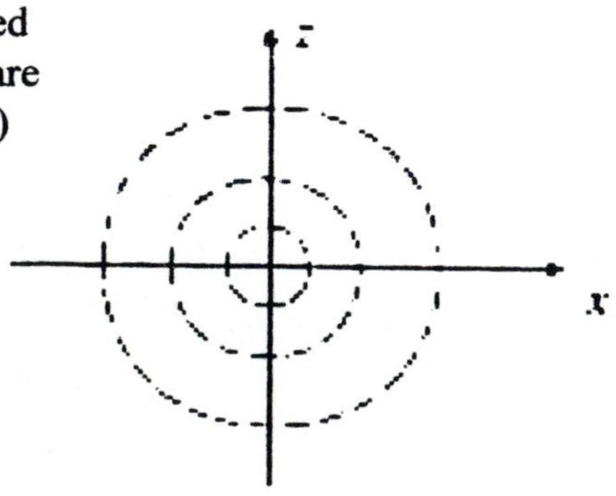

c. $5d_{x^2-y^2}$

Contours (dashed lines are nodes)

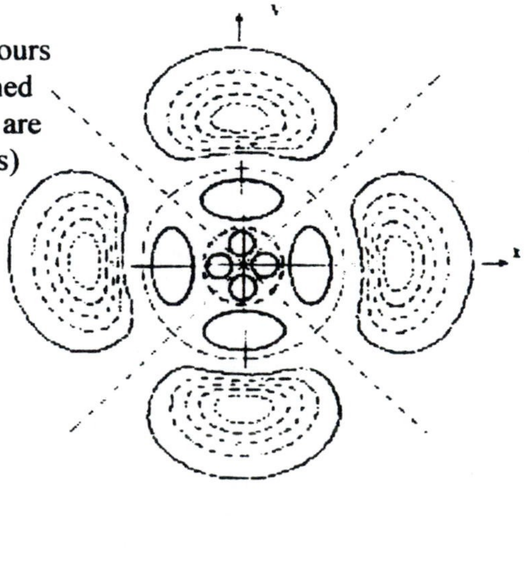

2-7 **a.**

5s

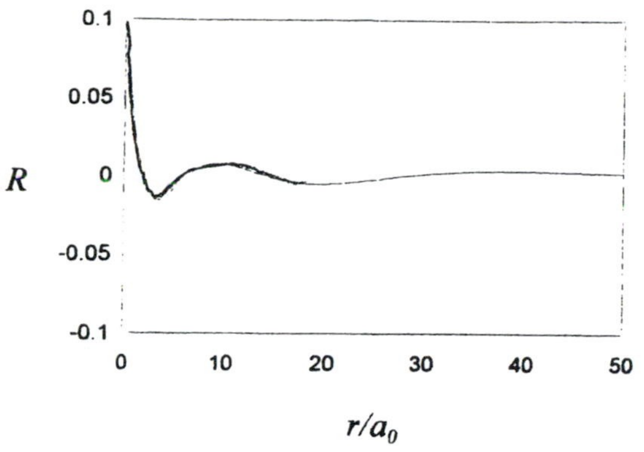

a.

$4d_{z^2}$

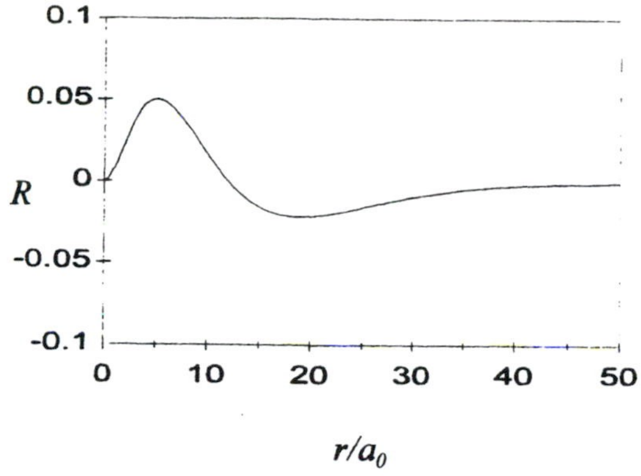

b.

5s

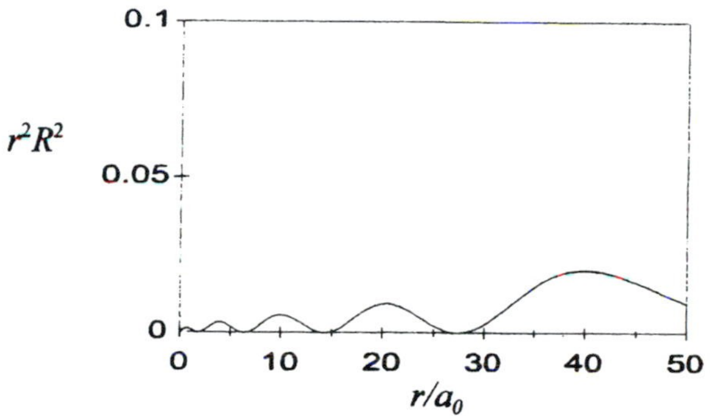

b.

$4d_{z^2}$

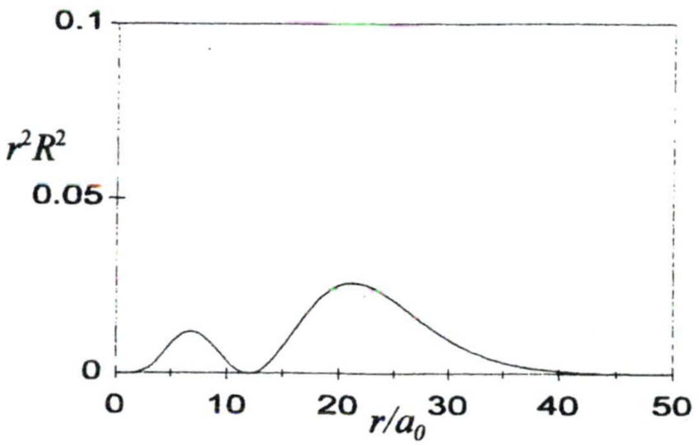

c.

5s

Contours (dashed lines are nodes)

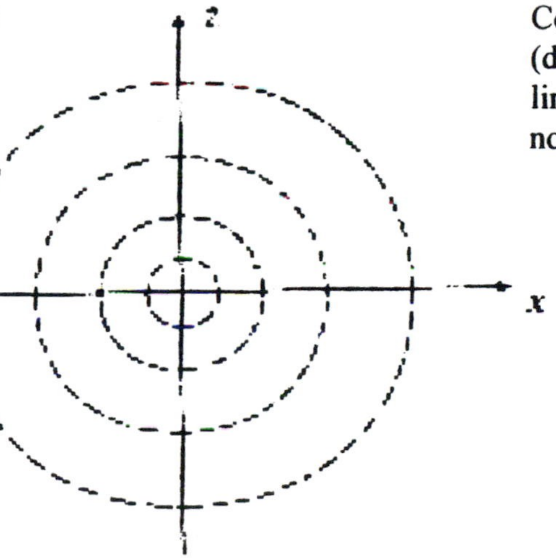

c.

$4d_{z^2}$

Contours (dashed lines are nodes)

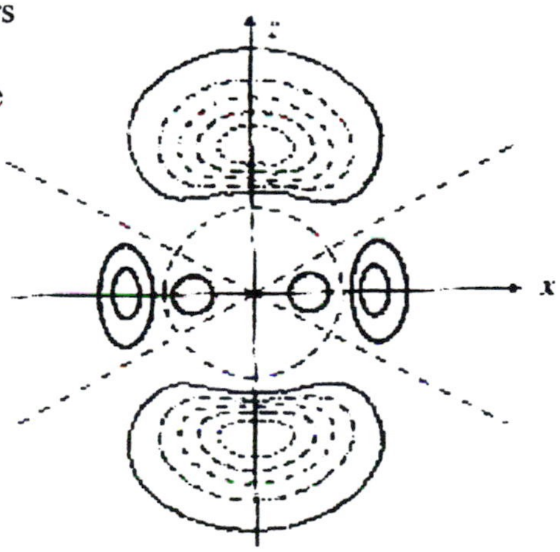

2-8 4 $f_{z(x^2-y^2)}$ orbital

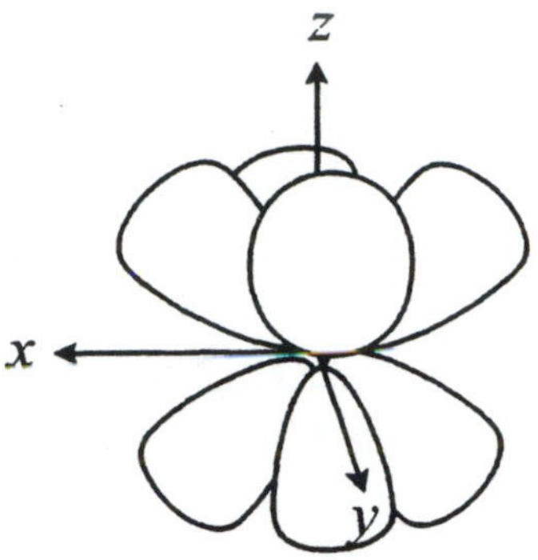

a. no spherical nodes

b. 3 angular nodes

c. The angular nodes are solutions for $z(x^2-y^2) = 0$. These solutions are $z = 0$ (*xy* plane), and the planes where $x = y$ and $x = -y$, both perpendicular to the *xy* plane

d. There are 8 lobes, 4 bisected by the *xz* plane and 4 by the *yz* plane. From the *z* axis, it looks like a $d_{x^2-y^2}$ orbital, but the node at the *xy* plane splits each lobe into two.

2-9 5 f_{xyz} orbital has the same general shape as the 4 $f_{z(x^2-y^2)}$, with the addition of a spherical node and rotation so the lobes are between the planes of the *xz* and *yz* axes rather than in the planes.

a. 1 spherical node

b. 3 angular nodes

c. The angular nodes are solutions for $xyz = 0$. These solutions are the three planes where $x = 0$ (*yz* plane), $y = 0$ (*xz* plane), and $z = 0$ (*xy* plane).

d. The diagram is similar to that in problem 2-8, with a spherical node added and rotated by 45° around the *z* axis. The 8 lobes, one in each octant of the coordinate system, are each cut by a spherical node added by the increase in principal quantum number compared to problem 2-8 for a total of 16 lobes.

2-10 a.

	n	l	m_l
5*d*	5	2	−2, −1, 0, 1, 2
4*f*	4	3	−3, −2, −1, 0, 1, 2, 3
7*g*	7	4	−4, −3, −2, −1, 0, 1, 2, 3, 4

b.

	n	l	m_l	m_s
3*d*	3	2	−2	±1/2
	3	2	−1	±1/2
	3	2	0	±1/2
	3	2	1	±1/2
	3	2	2	±1/2

10 possible combinations

2-11 a. Figure 2-12 and the associated text explain this. Electron-electron repulsion is minimized by placing each electron in a separate orbital when the levels are close enough in energy to allow it. At Cr, the second 4*s* electron has an energy higher than the lower five 3*d* electrons, therefore the configuration is $4s^1\,3d^5$.

b. Ti is $4s^2\,3d^2$, since both 4*s* electrons have energies below that of the 3*d* electrons at that *Z*. For ions, the 3*d* levels move down in energy, and are below the 4*s* levels for all transition metal M^{2+} ions, so Cr^{2+} is [Ar] $3d^4$.

2-12

a. V $1s^2\,2s^2\,2p^6\,3s^2\,3p^6\,4s^2\,3d^3$

b. Br $1s^2\,2s^2\,2p^6\,3s^2\,3p^6\,4s^2\,3d^{10}\,4p^5$

c. Ru^{3+} $1s^2\,2s^2\,2p^6\,3s^2\,3p^6\,3d^{10}\,4s^2\,4p^6\,4d^5$

d. Hg^{2+} $1s^2\,2s^2\,2p^6\,3s^2\,3p^6\,4s^2\,3d^{10}\,4p^6\,5s^2\,4d^{10}\,5p^6\,4f^{14}\,5d^{10}$

e. Sb $1s^2\,2s^2\,2p^6\,3s^2\,3p^6\,4s^2\,3d^{10}\,4p^6\,5s^2\,4d^{10}\,5p^3$

2-13 $3d^5$ is Mn^{2+}, $3d^2$ is Ti^{2+}

2-14

a. ↑ ↑ Π_e (A) ↑↓ _ Π_c (B) $\Pi_{(A-B)} = \Pi_e - \Pi_c$; Because Π_e is negative, A is favored regardless of Π_e and Π_c magnitudes.

b. ↑ ↑ ↑ $3\Pi_e$ (A) ↑↓ ↑ _ $\Pi_e + \Pi_c$ (B) $\Pi_{(A-B)} = 2\Pi_e - \Pi_c$; Again, A is favored

2-15

a.

	Z	$(1s^2)$ 1	$(2s^2\,2p^6)$ 0.85	$(3s^2\,3p^n)$ 0.35		*Z**	*r*
P	15 −	(2 × 1 +	8 × 0.85 +	4 × 0.35)	=	4.8	106 pm
S	16 −	(2 × 1 +	8 × 0.85 +	5 × 0.35)	=	5.45	102 pm
Cl	17 −	(2 × 1 +	8 × 0.85 +	6 × 0.35)	=	6.1	99 pm
Ar	18 −	(2 × 1 +	8 × 0.85 +	7 × 0.35)	=	6.75	98 pm

The size of the atoms decreases slightly as *Z* increases, even though the number of electrons in the atom increases, because *Z** increases and draws the electrons closer. Ar has the strongest attraction between the nucleus and the 3*p* electron, and the smallest radius.

b.

	Z	$(1s^2)$ 0.85	$(2s^2\,2p^n)$ 0.35		*Z**	*r*
O^{2-}	8 −	(2 × 0.85 +	7 × 0.35)	=	3.85	126 pm
F^-	9 −	(2 × 0.85 +	7 × 0.35)	=	4.85	119 pm
Na^+	11 −	(2 × 0.85 +	7 × 0.35)	=	5.85	116 pm
Mg^{2+}	12 −	(2 × 0.85 +	7 × 0.35)	=	6.85	86 pm

These values increase directly with *Z*, and parallel the decrease in ionic size. Increasing nuclear charge results in decreasing size for these isoelectronic ions, although the change between F^- and Na^+ is smaller than might be expected.

c. Cu $(1s^2)$ $(2s^2\ 2p^6)$ $(3s^2\ 3p^6)$ $(3d^{10})$ $(4s^1)$

$4s$ $S = 2 + 8 + (8\times0.85) + (10\times0.85) = 25.3$ $Z^* = 29 - 25.3 = 3.7$

$3d$ $S = 2 + 8 + (8\times1.00) + (9\times0.35) = 21.15$ $Z^* = 29 - 21.15 = 7.85$

The $3d$ electron has a nuch larger effective nuclear charge and is held more tightly; the $4s$ electron is therefore the first removed on ionization.

d. $(1s^2)$ $(2s^2\ 2p^6)$ $(3s^2\ 3p^6)$ $(3d^{10})$ $(4s^2\ 4p^6)$ $(4d^{10})$ $(4f^n)$

Ce $S = 2 + 8 + 8 + 10 + 8 + 10 = 46$

[Xe] $6s^2\ 4f^1\ 5d^1$ $4f^1$ $Z^* = 58 - 46 = 12$

Pr $S = 2 + 8 + 8 + 10 + 8 + 10 + (2\times0.35) = 46.7$

[Xe] $6s^2\ 4f^3$ $4f^3$ $Z^* = 59 - 46.7 = 12.3$

Nd $S = 2 + 8 + 8 + 10 + 8 + 10 + (3\times0.35) = 47.05$

[Xe] $6s^2\ 4f^4$ $4f^4$ $Z^* = 60 - 47.05 = 12.95$

The outermost electrons experience an increasing Z^*, and are therefore drawn in to slightly closer distances with increasing Z and Z^*.

2-16 **a.** Ca $Z^* = 20 - (2 + 8 + 8 \times 0.85 + 0.35) = 2.85$

Ga $Z^* = 31 - (2 + 8 + 18 \times 0.85 + 2 \times 0.35) = 5$

Based on Z^*, Ga should have the larger ionization energy. In fact, they are very similar, with Ca slightly larger, because Ga is losing a $4p$ electron, from a higher energy level.

b. Mg $Z^* = 12 - (2 + 8 \times 0.85 + 0.35) = 2.85$

Ca $Z^* = 20 - (2 + 8 + 8 \times 0.85 + 0.35) = 2.85$

Mg has a larger ionization energy in spite of the same Z^*, because the electron to be lost from Ca is farther from the nucleus.

c. Si $Z^* = 14 - (2 + 8 \times 0.85 + 3 \times 0.35) = 4.15$

P $Z^* = 15 - (2 + 8 \times 0.85 + 4 \times 0.35) = 4.80$

Z^* is calculated for the total number of electrons in the atom, because one electron is added to make the anion. In spite of the larger Z^* of P, Si has a larger electron affinity. The reaction of Si^-, $s^2p^3 \rightarrow s^2p^2 + e^-$, requires more energy than P^-, $s^2p^4 \rightarrow s^2p^3 + e^-$. The fourth electron in P^- is paired in one of the p orbitals. The pairing energy in P^- reduces the energy required for removal of an electron.

d. Mn^{2+} has the structure [Ar] $3d^5$. Transition metals always lose the s electrons first (see the explanation on pp. 41-43).

2-17 N $Z^* = 7 - (2 \times 0.85 + 4 \times 0.35) = 3.9$ IE = 1.402 MJ/mol r = 75 pm

P $Z^* = 15 - (2 \times 1 + 8 \times 0.85 + 4 \times 0.35) = 4.80$ IE = 1.012 MJ/mol r = 106 pm

As $Z^* = 33 - (2 \times 1 + 8 \times 1 + 18 \times 0.85 + 4 \times 0.35) = 6.3$ IE = 0.947 MJ/mol r = 120 pm

The effect of shielding alone is not sufficient to explain the changes in ionization energy. The other major factor is distance between the electron and the nucleus. Wulfsberg (*Principles of Descriptive Inorganic Chemistry*, Brooks/Cole, 1987) suggests that Z^*/r^2 correlates better, but As is still out of order:

	radius(pm)	Z^*/r^2
N	75	6.93×10^{-4}
P	106	4.27×10^{-4}
As	120	4.375×10^{-4}

2-18

		IE (kJ/mol)	
Cl^-	$p^6 - p^5$	349	$Z^* = 5.75$
Cl	$p^5 - p^4$	1251	$Z^* = 6.10$
Cl^+	$p^4 - p^3$	2300	$Z^* = 6.45$

In addition to the increase in the effective nuclear charge, the overall charge increases, making the second and third steps much more difficult (attractive rather than repulsive effects) and the radius decreases slightly with loss of each electron. All of these act to increase the ionization energy.

2-19

	Li >	Na >	K >	Rb
IE	5.39	5.14	4.34	4.18
Z^*	1.30	2.20	2.20	2.20
r(pm, covalent)	123	154	203	216
Z^*/r^2	8.59	9.28	5.34	4.72 (all $\times 10^{-5}$ pm^{-2})
r(pm, ionic)	90	116	152	166
Z^*/r^2	16.0	16.3	9.52	7.98 (all $\times 10^{-5}$ pm^{-2})

The Z^*/r^2 function explains the order, except for lithium, which seems to require more energy for removal of the electron than predicted. Apparently the very small size and small number of electrons on lithium result in the electron being held more tightly than in the other alkali metals. The Z^*/r^2 function also predicts larger differences between the IE values than are observed.

2-20 C^+ $(1s^2)(2s^2\,2p^1)$

$$Z^* = 6 - (2 \times 0.85 + 2 \times 0.35) = 3.6$$

B $(1s^2)(2s^2\,2p^1)$

$$Z^* = 5 - (2 \times 0.85 + 2 \times 0.35) = 2.6$$

The energies change by a factor of three, but Z^* changes only by 38%. Based on the data in Table 2-8 and the relative ionic and covalent sizes in problem 19, C^+ has a radius of about 58 pm, B a radius of about 82 pm. Z^*/r^2 values are then 10.7×10^{-4} and 3.8×10^{-4}, a ratio of 2.8. The ionization energies have a ratio of 2.99. This function of size and effective charge explains this pair quite well.

2-21 **a.** Fe (7.87 eV) > Ru (7.37 eV) They have the same Z^* (6.25), but Ru is larger, so Z^*/r^2 is smaller than for Fe.

b. P (10.486 eV) > S (10.36 eV) Z^* is smaller for P (4.8) than for S (5.45), but S has one electron paired in the $3p$ level, which increases its energy and makes it easier to remove.

c. Br (11.814 eV) > K (4.341) Z^* for K is 2.2; for Br it is 7.6, a very large difference. K is also nearly twice as large as Br. Both factors contribute to the difference in IE.

d. N (14.534) > C (11.260) Increasing Z^* (3.9 for N, 3.25 for C) and decreasing size (75 pm for N, 77 pm for C) both work in the same direction.

e. Cd (8.993) > In (5.786) In starts a new $5p$ subshell, so the last electron is easily removed in spite of a larger Z^* (2.30 for In, 1.65 for Cd).

2-22 a. S (2.077 eV) has a smaller EA than Cl (3.617 eV) because Cl has a larger Z^* (6.1 vs. 5.45) and a slightly smaller radius. Both increase the attractive power for an electron.

b. I (3.059) has a smaller EA than Br (3.365) because it is larger, with the same Z^* (7.6).

c. B (8.298) has a smaller IE than Be (9.322) because it is starting a new p subshell.

d. S (10.360) has a smaller IE than P (10.486) because S^- is losing one of a pair of p electrons (similar to N and O in problem 21c).

e. Cl (12.977) has a smaller IE than F (17.422) because it is larger and has a smaller Z^*. Z^*/r^2 values do not work well for these atoms, predicting more than a 2:1 ratio. (Z^*/r^2 values are 4.49 and 10.0×10^{-4}, respectively.)

2-23 a. The maximum at Mg comes at a completed subshell ($3s^2$) and the minimum at Al is at the first electron of a new subshell ($3p$), increasing the energy and making removal easier. The maximum at P is at a half-filled subshell ($3p^3$) and the minimum at S is at the fourth electron, which must pair with one of the others, which also raises the energy of the electron and makes its removal easier.

b. The reasons for the minima and maxima in the electron affinity graph are the same as in the ionization energy graph. The maxima and minima are shifted by one in the two graphs because the reactants in the process defining electron affinity have a negative charge, one more electron than a neutral atom. For example, minima occur for ionization energy at Al ([Ne]$3s^23p^1$ $\rightarrow$) and for electron affinity at Mg ([Ne]$3s^23p^1$ $\rightarrow$); Al and Mg^- have identical electron configurations.

2-24 The Bohr equation (p. 17) predicts that the energy levels of 1-electron systems should be proportional to Z^2. $Z^2 = 4$ for He^+ and 9 for Li^{2+}, so ratios of the ionization energies to that of H are 4:1 and 9:1.

2-25 In both the transition metals and the lanthanides (problem 15d), the gradual change in Z^* pulls the outer electrons closer. The increase in Z^* is 0.65 for each unit increase in atomic number, so the increase in attraction is relatively small and the change in radius must also be small.

2-26 **a.** $Se^{2-} > Br^- > Rb+ > Sr^{2+}$ These ions are isoelectronic, so the sizes are directly dependent on the nuclear charges. For example, Sr^{2+} has the greatest nuclear charge and is the smallest ion.

b. $Y^{3+} > Zr^{4+} > Nb^{5+}$ These ions are also isoelcctronic, so the increasing nuclear charge results in decreasing size.

c. $Co^{4+} < Co^{3+} < Co^{2+} < Co$ The smaller number of electrons with constant nuclear charge results in a smaller size.

2-27 Zr through Pd have the electron configurations shown below:

40	Zr	$5s^2\ 4d^2$
41	Nb	$5s^1\ 4d^4$
42	Mo	$5s^1\ 4d^5$
43	Tc	$5s^2\ 4d^5$
44	Ru	$5s^1\ 4d^7$
45	Rh	$5s^1\ 4d^8$
46	Pd	$4d^{10}$

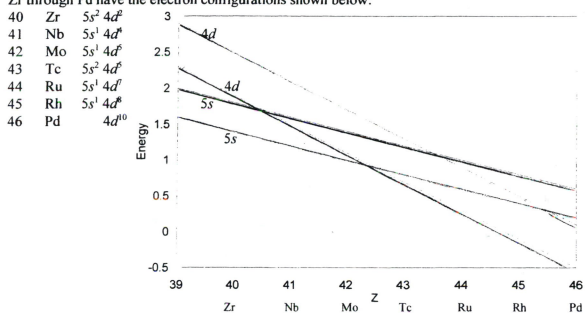

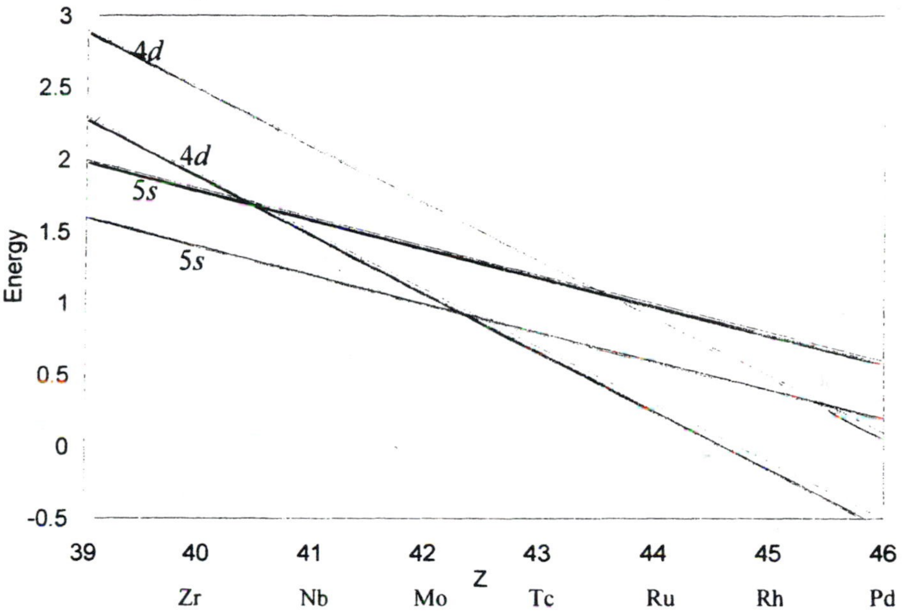

The lower *d* line crosses the upper *s* line between 40 and 41, the upper *d* line crosses the upper *s* line between 43 and 44, and the upper *d* line crosses the lower *s* line between 45 and 46. This graph fits the experimental configurations well.

2-28 http:\\www.orbital.com has 3D representations of orbitals of all kinds. Other sites can be found by doing a search for f orbitals or g orbitals. The authors would be interested in learning of the sites you and your students find.

Chapter 3
Simple Bonding Theory

3-1 **a.** Structures a and b are more likely than c, because the negative formal charge is on the electronegative S. In c, the electronegative N has a positive charge.

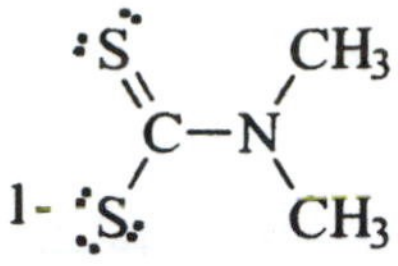

a

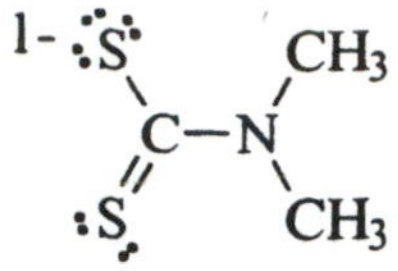

b

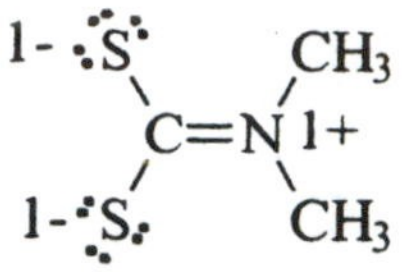

c

b. The same structures fit $OSCN(CH_3)_2^-$. The structure with 1- formal charge on O is most likely, since O is the most electronegative atom in the ion.

3-2 **a.**

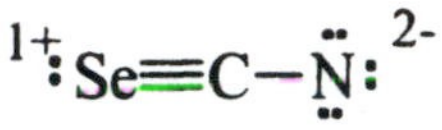

The formal charges are large, but match electronegativity.

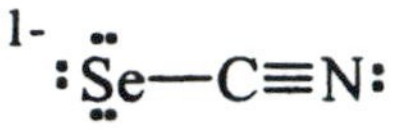

Negative formal charge of 1− on Se, a low electronegativity atom.

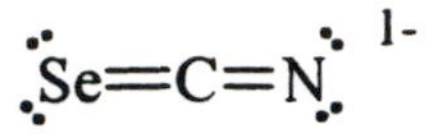

Negative formal charge on N, the most electronegative atom. Best of the three.

b. b is better than a, because the formal charge is on the more electronegative O.

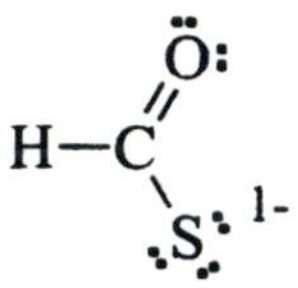

a

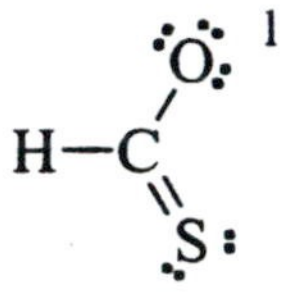

b

c. a and b are better than c, because one of the formal charges is on the more electronegative O.

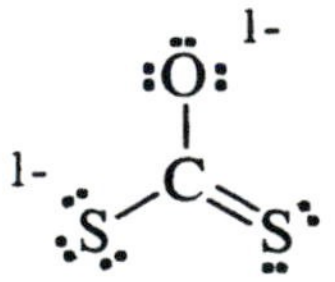

a

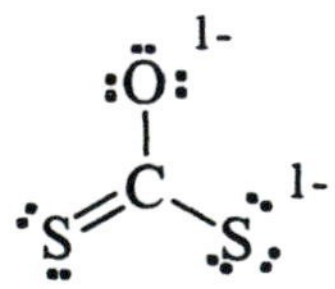

b

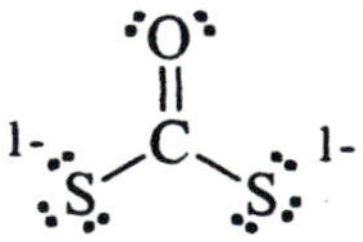

c

3-3 NSO^-: a has 2− formal charge on N, 1+ on S. Large charge, not very likely. b has 1− formal charges on N and O, 1+ on S, and is a better structure.

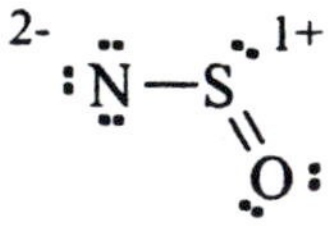

a

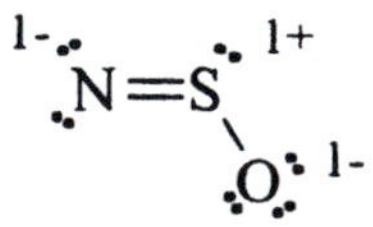

b

SNO^-: a has 1− formal charge on S. Not very likely, doesn't match electronegativity. b has 1− formal charge on O, and is a better structure.

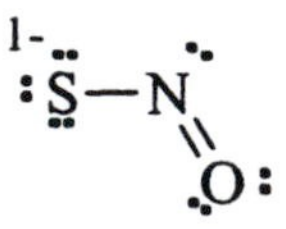

a

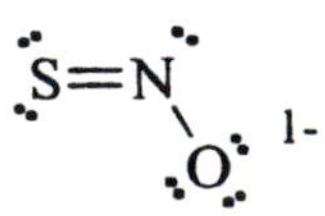

b

Overall, $S{=}N{-}O^-$ is better based on formal charges, since O is the most electronegative atom in the ion.

3-4

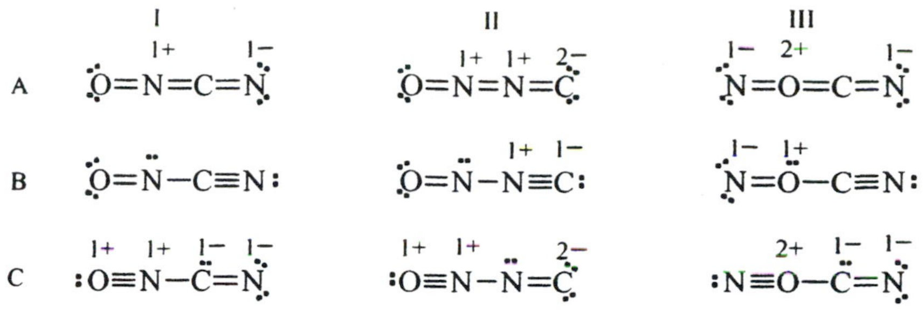

Structure IB is best by the formal charge criterion, with no formal charges, and is expected to be the most stable. None of structures II or III are good, with unlikely charges (by electronegativity arguments) or large charges.

3-5 IF_3^{2-} requires 12 electrons on I, and has three lone pairs and three bonds. Overall, the ion is predicted to be T-shaped, with bond angles slightly less than 90°.

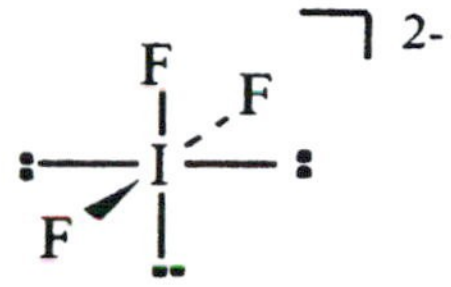

3-6 **a.** AsH_3 should have the smallest angle, since it has the largest central atom. This minimizes the bond pair-bond pair repulsions and allows a smaller angle. Arsenic is also the least electronegative central atom, allowing the electrons to be drawn out farther and lowering the repulsions further. Actual angles: AsH_3 = 91.8°, PH_3 = 93.8°, NH_3 = 106.6°.

b. O_3^+ (unknown angle) should be near the trigonal angle, with a slightly larger angle based on the odd electron spending some time on the center O. O_3 (116.8°) has a slightly smaller angle than O_3^+, with a pair of electrons on the center O. O_3^- (100°) is near a tetrahedral angle, with a slightly smaller angle than 109.5° due to the lone pair and odd electron in the central O. Therefore, O_3^- has the smallest angle.

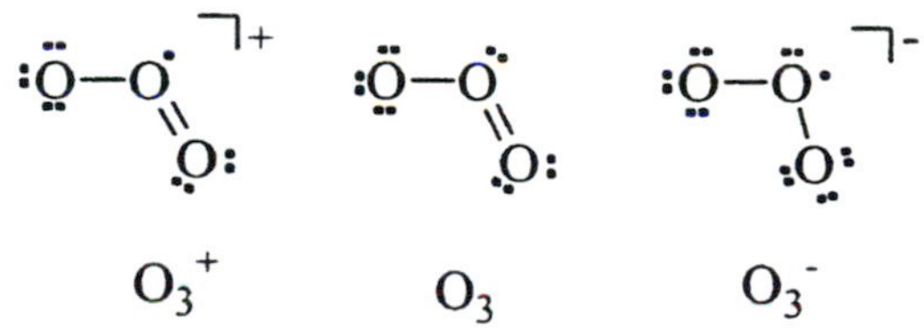

c. Cl is larger than F, and F is more electronegative and should pull the electrons farther from the S, so the F-S-F angle should be smaller. This is consistent with the experimental data: the F-S-F angle in OSF_2 is 92.3° and the Cl-S-Cl angle in $OSCl_2$ is 96.2°

d. NO_2^- has rather variable angles (115° to 132°) in different salts. The sodium salt(115.4°) has a slightly smaller angle than O_3 (116.8°). The N-O electronegativity difference should pull electrons away from N, reducing the bp-bp repulsion and the angle.

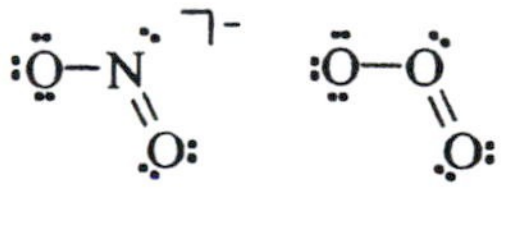

e. BrO_3^- (104°) has a slightly smaller angle than ClO_3^- (107°), since it has a larger central atom, as in problem 3-6a. In addition, the larger electronegativity of Cl holds the electrons closer and increases *bp-bp* repulsion.

3-7 PCl_3Br_2 requires 10 electrons around P. The location of the Cl and Br atoms is uncertain, but comparison with PCl_2F_3 suggests that two Cl atoms are in axial positions. The Br atoms require more space, both on electronegativity and size arguments. As a result, they should occupy equatorial positions.

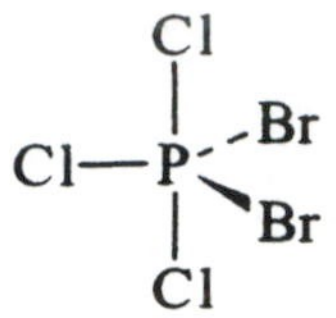

3-8 **a.** $SeCl_4$ requires 10 electrons around Se. The lone pair of electrons in an equatorial position of a trigonal bipyramid distorts the shape by bending the axial Cl's back.

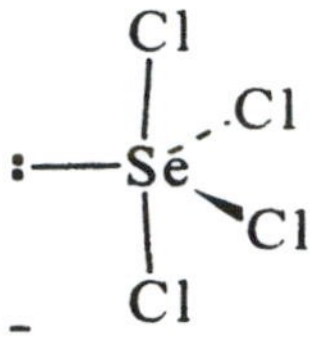

b. I_3^- requires 10 electrons around the central I and is linear.

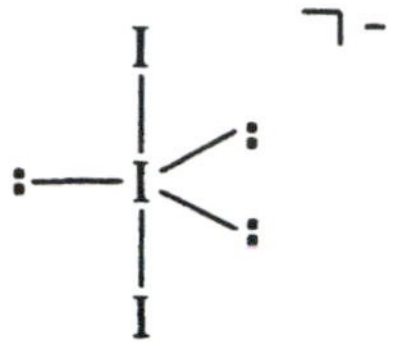

c. $PSCl_3$ is nearly tetrahedral.

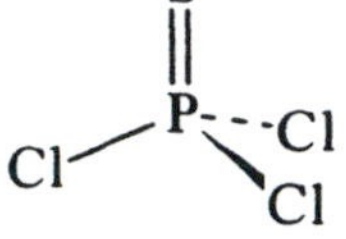

d. IF_4^- has 12 electrons around I and has a square planar shape.

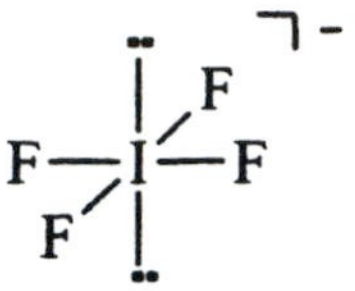

e. PH_2^- has a bent structure, with two lone pairs.

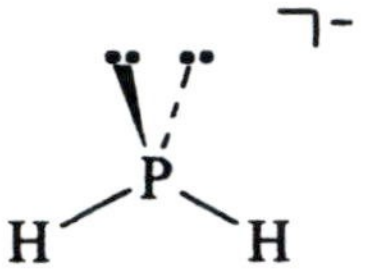

f. TeF_4^{2-} has 12 electrons around Te, with a square planar shape.

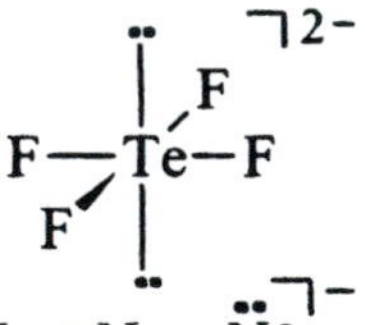

g. N_3^- is linear, with two double bonds (best structure) or one single bond and a triple bond as resonance structures.

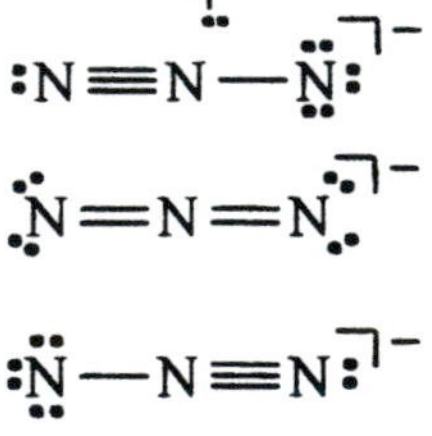

h. $SeOCl_4$ has a distorted trigonal bipyramidal shape with the extra repulsion of the double bond placing oxygen in an equatorial position.

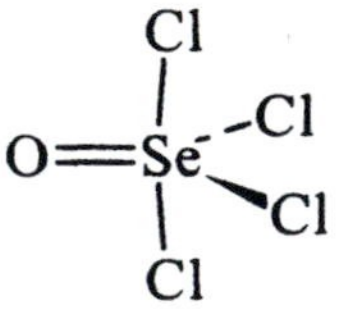

i. PH_4^+ is tetrahedral.

j. NO^- must be linear, and has a double bond.

3-9 **a.** ICl_2^- has 10 electrons around I, and a linear shape.

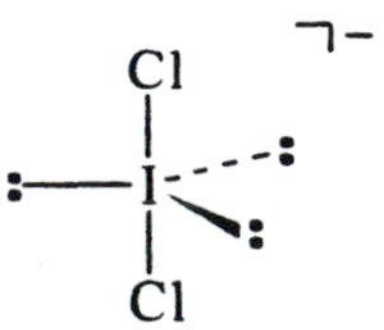

b. H_3PO_3 has a distorted tetrahedral shape.

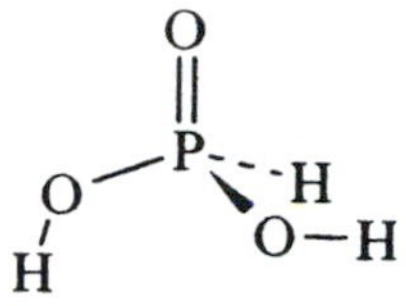

c. BH_4^- is tetrahedral.

d. $POCl_3$ is a distorted tetrahedron.

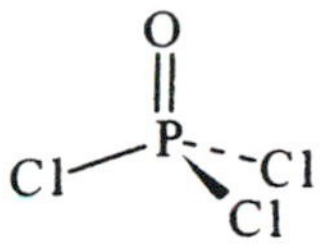

e. IO_4^- has a tetrahedral shape with three double bonds in its most favorable resonance structure to reduce the formal charge on I.

f. $IO(OH)_5$ has the oxygens arranged octahedrally, with hydrogens on five of the six oxygens.

g. $SOCl_2$ is trigonal pyramidal, with one lone pair and some double bond character in the S–O bond.

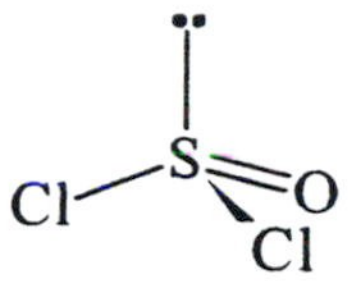

h. $ClOF_4^-$ is a square pyramid. The double bonded O and the lone pair compete, with little difference between their repulsive forces.

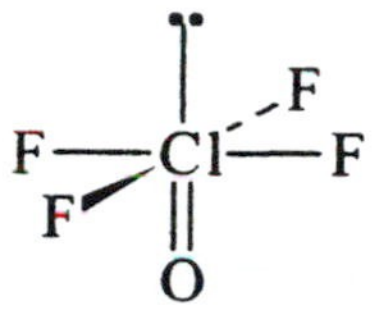

i. XeO_2F_2 has 10 to 14 electrons around Xe, depending on the number of double bonds drawn. F–Xe–F is nearly linear (174.7°), with the two oxygens and a lone pair in a trigonal planar configuration. Formal charges favor double bond character in the Xe–O bonds. The O–Xe–O angle is narrowed to 105.7° by *lp-bp* repulsion.

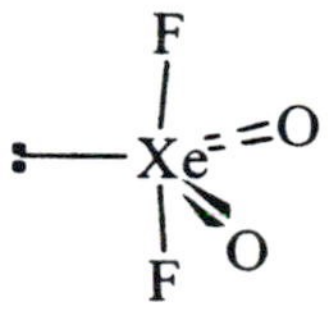

j. $ClOF_2^+$ is trigonal pyramidal with one lone pair and double bond character in the Cl–O bond.

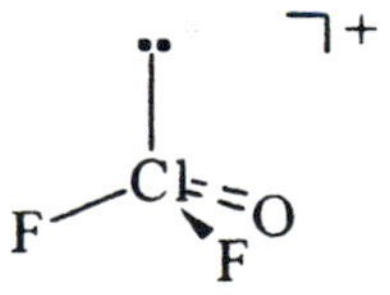

3-10 a. SOF_6 is nearly octahedral around the S.

b. POF_3 has a distorted tetrahedral shape, with F–P–F angles of 101°.

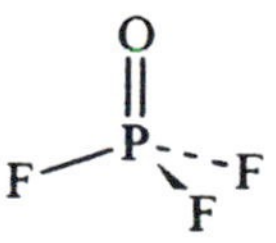

c. ClO_2 is an odd electron molecule, with a bent shape, partial double bond character, and an angle of 117.5°.

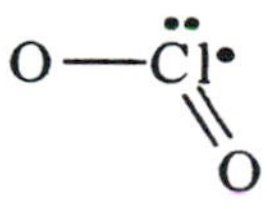

d. NO_2 is another odd electron molecule, bent, with partial double bond character and an angle of 134.25°. This is larger than the angle of ClO_2 because there is only one odd electron on N, rather than the one pair and single electron of ClO_2.

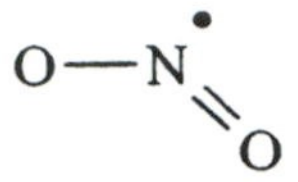

e. $S_2O_4^{2-}$ has SO_2 units with an angle of about 30° between their planes, in an eclipsed conformation.

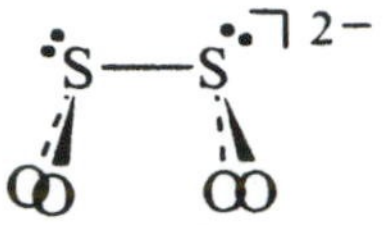

f. N_2H_4 has a trigonal pyramidal shape at each N, and a *gauche* conformation. One lone pair on each N.

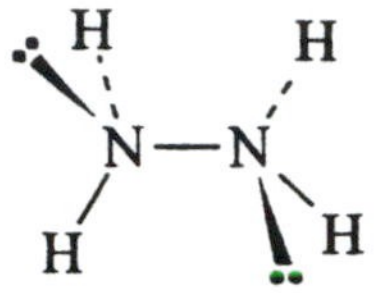

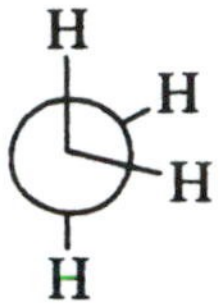

3-11 a. N_3^- is linear, with two double bonds. O_3 is bent, with one double bond and a lone pair on the central O caused by the extra pair of electrons.

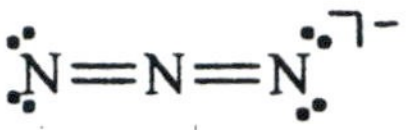

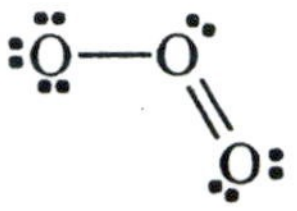

b. Adding an electron to O_3 decreases the angle, as the odd electron spends part of its time on the central O, making two positions for electron repulsion. See Problem 3-6b.

3-12 a. $VOCl_3$ has a distorted tetrahedral shape, with Cl-V-Cl angles of 111°, and Cl-V-O angles of 108°.

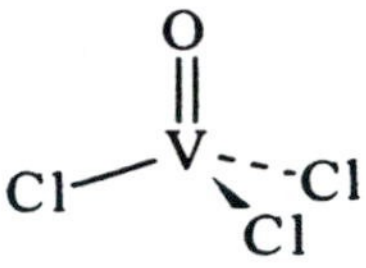

b. PCl_3 has a trigonal pyramidal shape with Cl-P-Cl angles of 100°.

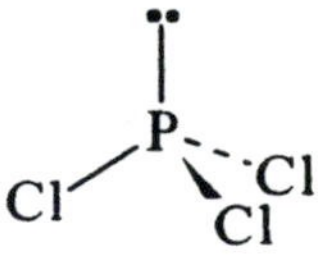

c. SOF_4 has a distorted trigonal bipyramidal shape. The axial fluorine atoms are nearly linear with the S atom; the equatorial F-S-F angle is 100°.

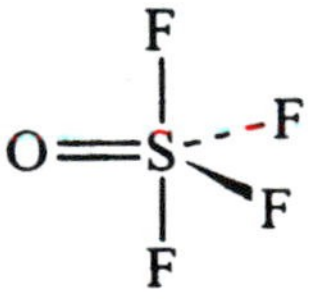

d. ClO_2^- is bent, with two lone pairs on the Cl. The angle is variable, but near 108°

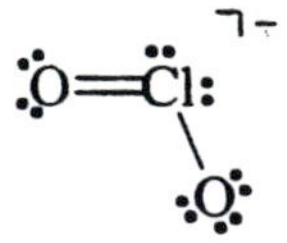

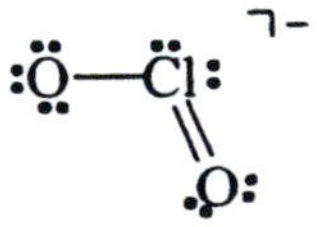

e. ClO_3^- has a trigonal pyramidal shape, with one lone pair and three principal resonance structures and a bond angle of 107° for O-Cl-O.

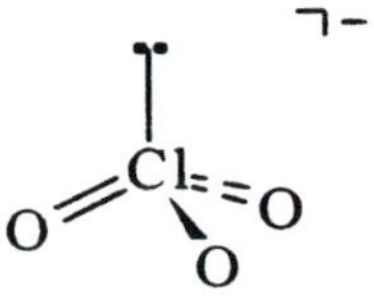

f. P_4O_6 is described in the problem. Each P has one lone pair, each O has two.

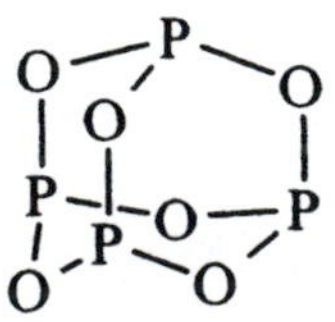

3-13 The bond angles increase as the groups become bulkier (H < CH_3 < SiH_3 < GeH_3). The last two are planar, and the angles between the groups are at a maximum. Si and Ge also have *d* orbitals that can form π bonds with the axial N *p* orbital. (D. W. H. Rankin and H. E. Robertson, *J. Chem. Soc. Dalton*, **1987**, 785)

3-14 See problems 3-6e and 3-12. All the halate ions are trigonal pyramids; as the central atom increases in size, the bonding pairs are farther from the center, and the lone pair forces a smaller angle. The decreasing electronegativity Cl > Br > I of the central atom also allows the electrons to be pulled farther out, reducing the *bp-bp* repulsion.

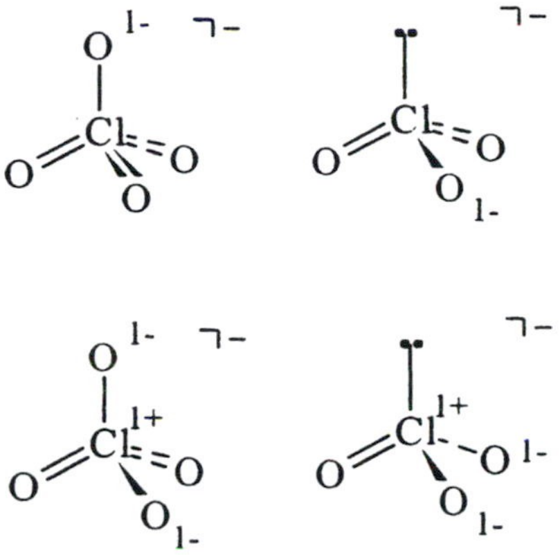

3-15 ClO_4^- will have slightly more double bond character than ClO_3^-. If the formal charge on Cl is zero in each, ClO_4^- has $1\frac{3}{4}$ bond for each Cl–O and ClO_3^- has $1\frac{2}{3}$ bond. Allowing 1+ on Cl, ClO_4^- has $1\frac{1}{2}$ bond for each and ClO_3^- has $1\frac{1}{3}$. Alternative descriptions of ions like these with more "ionic" character in the bonds are comparable to the SO_3^{2-} and SO_4^{2-} structures described in Chapter 5 (p. 161).

3-16 **a.** PH_3 has a smaller bond angle than NH_3, about 93°. The larger central atom reduces the repulsion between the bonding pairs.

b. H_2Se has a structure like water, with a bond angle near 90°. The larger central atom increases the distance between the S-H bonding pairs and reduces their repulsion, resulting in a smaller angle.

c. SeF_4 has a lone pair at one of the equatorial positions of a trigonal bipyramid, and bond angles of about 110° (equatorial) and 169° (axial). Teeter-totter shape.

d. PF_5 has a trigonal bipyramidal structure.

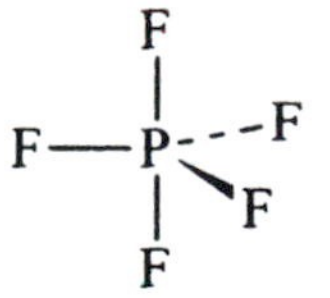

e. ICl_4^- has the four Cl's in a square planar shape, with lone pairs in the axial positions.

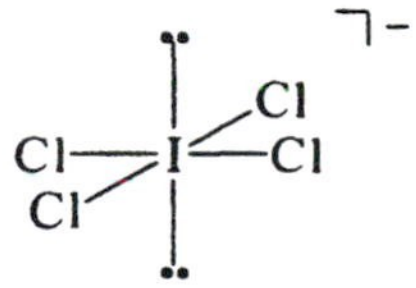

f. XeO_3 has a trigonal pyramidal shape, similar to NH_3, but with Xe–O double bonds.

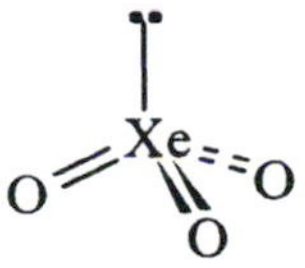

g. NO_3^- is a planar trigonal ion, with three resonance structures and 1 1/3 bond order to each oxygen.

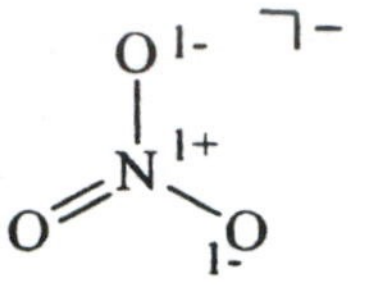

h. $SnCl_2$ has a bond angle of 95° in the vapor phase, smaller than the trigonal angle. As a solid, it forms polymeric chains with bridging Cl's and bond angles near 80°.

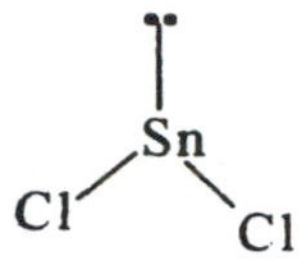

i. PO_4^{3-} is tetrahedral. Bond order is near 1 1/3.

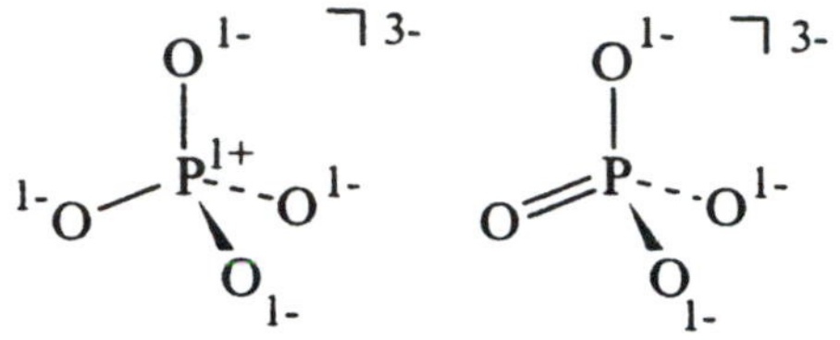

j. SF_6 is octahedral, with all single bonds.

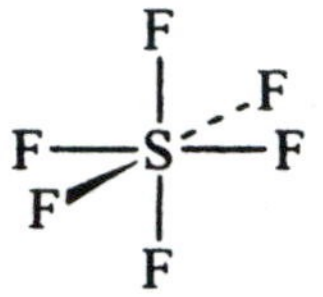

k. IF_5 is square pyramidal, with slight distortion away from the lone pair.

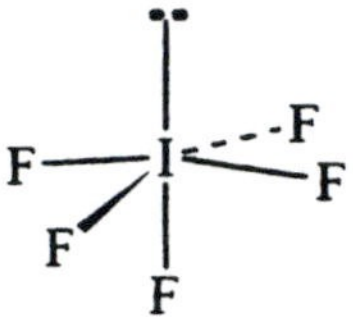

l. ICl_3 is unstable in the gas phase, forming ICl and Cl_2. In the solid state, it forms I_2Cl_6 dimers, with two Cl atoms bridging between the I atoms. The ICl_3 molecule should be T-shaped, with. bond angles slightly less than 90° due to repulsion by the two lone pairs.

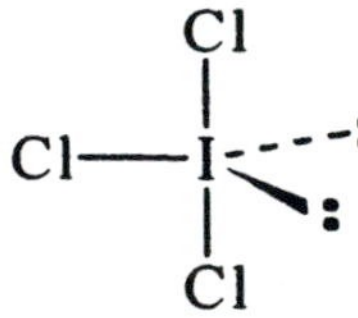

m. $S_2O_3^{2-}$ is nearly tetrahedral, with bond orders near 1.5, as in structure (c).

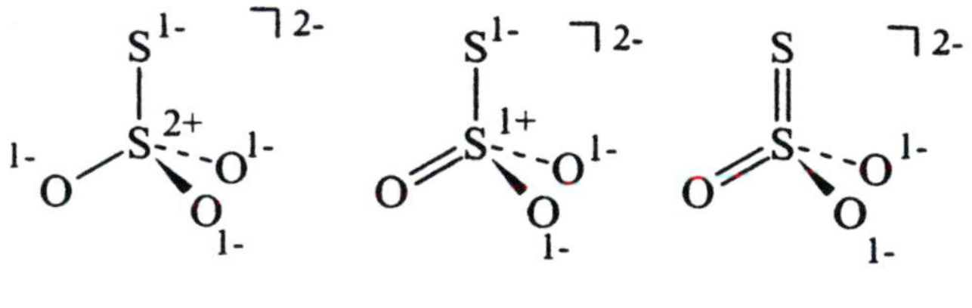

(a) (b) (c)

n. BF_2Cl is trigonal planar, with some small double bonding contribution, and < FBCl larger than <FBF.

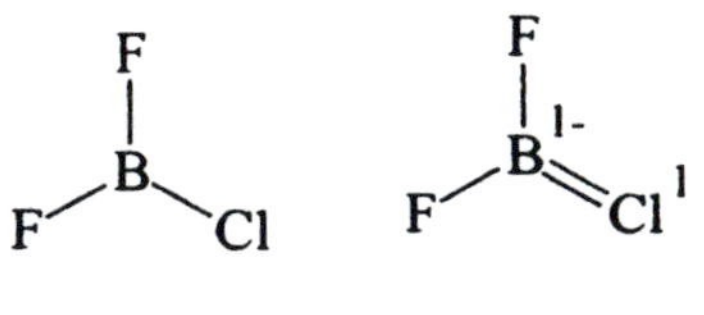

3-17 Polar molecules in 3-16: PH_3 H_2Se SeF_4 XeO_3 $SnCl_2$ IF_5 ICl_3 $S_2O_3^{2-}$ BF_2Cl

3-18 CO has about 76 kJ contribution to its bond energy because of the electronegativity difference between C and O; attraction between the slightly positive and negative ends strengthens the bonding. Although this is not a complete explanation, it covers most of the difference between CO and N_2. In spite of its high bond energy, N_2 is thought by some to have some repulsion in its sigma bonding because of the short bond distance.

3-19 **a.** CF_3 has a greater attraction for electrons than CH_3, so the P in $PF_2(CF_3)_3$ is more positive than the P in $PF_2(CH_3)_3$. This draws the F atoms in slightly, so the P—F bonds are shorter in $PF_2(CF_3)_3$ (160.1 pm vs. 168.5 pm).

b. Al—O—Al could have an angle near 109°, like water, or could have double bonds in both directions and a nearly linear structure. In fact, the angle is about 140°. The single-bonded picture is more probable; the high electronegativity of O compared to Al draws the bonding pairs closer, opening up the bond angle. A Lewis structure with zero formal charges on all atoms can be drawn for this molecule with four electrons on each Al.

c. CAl_4 is tetrahedral. Again, a Lewis structure with zero formal charges can be drawn with four electrons on each Al.

3-20 The electronegativity differences are given in parentheses.

a.	C-N	N is negative (0.522)
b.	N-O	O is negative (0.544)
c.	C-I	C is negative (0.185)
d.	O-Cl	O is negative (0.741)
e.	P-Br	Br is negative (0.432)
f.	S-Cl	Cl is negative (0.280)

The overall order of polarity is O-Cl > N-O > C-N > P-Br > S-Cl > C-I.

3-21 **a.** PCl_5 has 10 electrons around P, using the 3*d* orbitals in addition to the usual 3*s* and 3*p*. N is too small to allow this structure. In addition, N would require use of the 3*s*, 3*p*, or 3*d* orbitals, but they are too high in energy to be used effectively.

b. Similar arguments apply, with O too small and lacking in accessible higher energy orbitals.

3-22 **a.** The H-O bond of methanol is more polar than the H-S bond of methyl mercaptan. As a result, hydrogen bonding holds the molecules together and requires more energy for vaporization. The larger molecular weight of methyl mercaptan has the opposite effect, but the hydrogen bonding has a stronger influence.

b. CO and N_2 have the same molecular weights, so the polarity of CO is the only factor helping hold it in the solid or liquid states.

c. The *ortho* isomer of hydroxybenzoic acid can form intramolecular hydrogen bonds, while the *meta* and *para* isomers tend to form dimers and larger aggregates in their hydrogen bonding. As a result, the *meta* and *para* isomers have higher melting points. (*ortho*, 159°; *meta*, 201.3°; *para*, 214-5°)

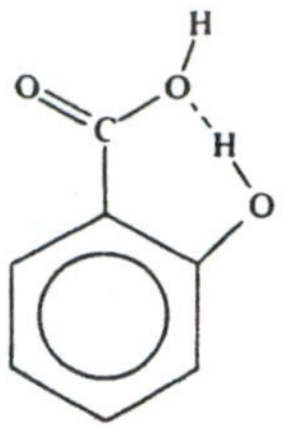

d. The London (dispersion) forces between atoms increase with the number of electrons, so the noble gases with larger Z have larger interatomic forces and higher boiling points.

e. Acetic acid can form hydrogen-bonded dimers in the gas phase, so the total number of particles in the gas is half the number expected by using the ideal gas law.

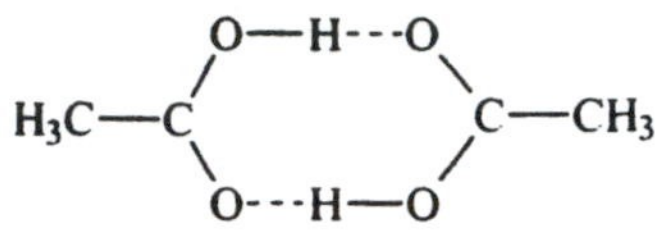

f. Acetone has a negative carbonyl oxygen; chloroform has a positive hydrogen, due to the negative character of the Cl's. As a result, there is a stronger attraction between the different kinds of molecules than between molecules of the same kind, and a resulting lower vapor pressure. This is an unusual case of hydrogen bonding, with no H-N, H-O, or H-F bond involved.

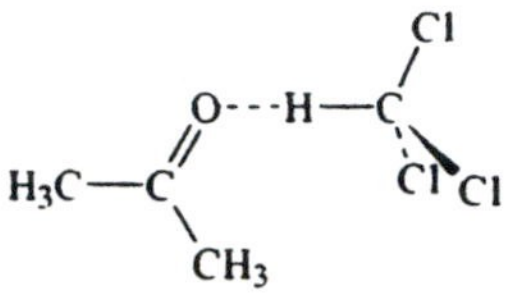

3-23 Using electronegativity values from Table 3-3, the following calculations can be made.

Molecul e	*Ato m*	*Group number*	*Unshared electrons*	$2\left(\frac{\chi_A}{\chi_A + \chi_B}\right)$	*Number of bonds*	*Formal charge*
C≡O	C	4	2	$2\left(\frac{2.544}{2.544 + 3.61}\right) = 0.83$	3	-0.49
	O	6	2	$2\left(\frac{3.61}{2.544 + 3.61}\right) = 1.17$	3	0.49
$N{=}O^-$	N	5	4	$2\left(\frac{3.066}{3.066 + 3.61}\right) = 0.92$	2	-0.84
	O	6	4	$2\left(\frac{3.61}{3.066 + 3.61}\right) = 1.08$	2	-0.16
H-F	H	1	0	$2\left(\frac{2.300}{2.300 + 4.193}\right) = 0.71$	1	0.29
	F	7	6	$2\left(\frac{4.193}{2.300 + 4.193}\right) = -1.29$	1	-0.29

Surprisingly, CO is more polar than HF and NO^- is intermediate, with C and N the negative atoms in CO and NO^-.

3-24 $I(CF_3)Cl_2$ is roughly T-shaped, with the two Cl atoms opposite each other and the CF_3 group and two lone pairs in the trigonal plane. The experimental Cl-I-C angles are 88.7° and 82.9°, smaller than the 90° expected if there were no extra repulsion from the lone pairs. Repulsion between the lone pairs and the larger CF_3 group put them in the trigonal plane, where there is more room.

3-25 C_3O_2 has the linear structure O=C=C=C=O, with zero formal charges.

N_5^+ with the same electronic structure has formal charges of −1, +1, +1, +1, −1, unlikely because three positive charges are adjacent to each other. Changing to N=N=N−N≡N results in formal charges of −1, +1, 0, +1, 0, a more reasonable result with an approximately trigonal angle in the middle. With triple bonds on each end, the formal charges are 0, +1, −1, +1, 0 and a tetrahedral angle. Some contribution from this would reduce the bond angle.

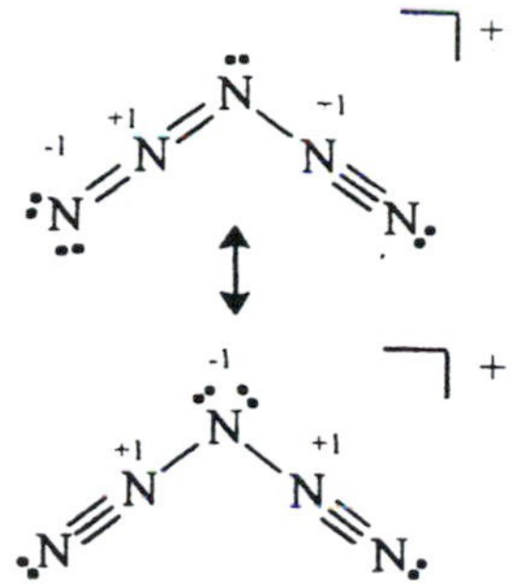

OCNCO$^+$ can have the structure O≡C−N−C≡O, with formal charges of +1, 0, −1, 0, +1 and two lone pairs on the central N. This would result in an even smaller angle in the middle, but has positive formal charges on O, the most electronegative atoms. O=C=N−C≡O has a formal charge of +1 on the final O. Resonance would reduce that formal charge, making this structure and a trigonal angle more likely. From Seppelt: "the markedly higher electronegativity of the nitrogen atom with respect to the central atom in C_3O_2, which leads to a higher localization of electron density in the sense of a nonbonding electron pair." He also mentions two lone pairs on N.

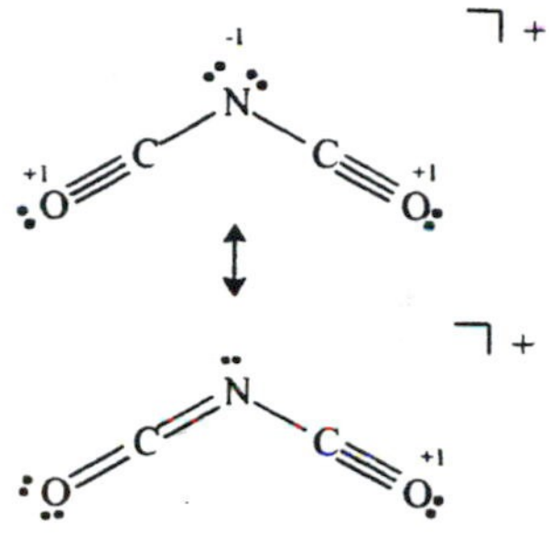

Therefore, the bond angles should be OCCCO > OCNCO$^+$ > N_5^+. Literature values are 180°, 130.7°, and 108.3 to 112.3° (calculated), respectively.

3-26 **a.** O is more electronegative than N and can draw the electrons away from the S. This in turn draws electrons toward the S, and shortens the S—Cl bonds. Shorter bonds result in larger bond angles, so $SOCl_2$ should have a larger Cl-SI-Cl bond angle. Experimentally, $NSCl_2^-$ has a short N—S bond (143.6 pm, compared with the O—S length of 144.6 pm of $OSCl_2$) and long S—Cl bonds (242.3 pm, compared with the S—Cl length of 207.6 pm of $OSCl_2$). The Cl-S-Cl angle is 93.3° and the N-S-Cl angle is 112.8°. The $OSCl_2$ Cl-S-Cl angle is 96.2°, and the Cl-S-O angle is 107.3°. ($OSCl_2$ data from I. Hargittai, *Magy. Chem. Foly.*, **1968**, *74*, 596.)

b. As described above, the $NSCl_2$ has longer S—Cl bonds.

3-27 Bond angles and distances:

	Steric Number	C—F (pm)	< FCF (°)	F—F (pm)
$F_2C{=}CF_2$	3	133.6	109.2	218
F_2CO	3	131.9	107.6	216
CF_4	4	131.9	109.5	216
F_3CO^-	4	139.2	101.3	215

The differences between these molecules are very subtle. The LCP model views the F ligands as hard objects, tightly packed around the central C in these examples. In this approach, the F-F distance remains nearly constant while the central atom moves to minimize repulsions.

Chapter 4
Symmetry and Group Theory

4-1 **a.** Ethane in the staggered conformation has 2 C_3 axes (the C−C line), 3 perpendicular C_2 axes bisecting the C−C line, in the plane of the two C's and the H's on opposite sides of the two C's. No σ_h, 3 σ_d, i, S_6. Overall, a $\boldsymbol{D_{3d}}$ molecule.

b. Ethane in eclipsed conformation has two C_3 axes (the C−C line), three perpendicular C_2 axes bisecting the C−C line, in the plane of the two C's and the H's on the same side of the two C's. Mirror planes include σ_h and 3 σ_d. Overall, a $\boldsymbol{D_{3h}}$ molecule.

c. Chloroethane in the staggered conformation has only one mirror plane, through both C's, the Cl, and the opposite H on the other C. Overall, a $\boldsymbol{C_s}$ molecule.

d. 1,2-dichloroethane in the *trans* conformation has a C_2 axis perpendicular to the C−C bond and perpendicular to the plane of both Cl's and both C's, a σ_h plane through both Cl's and both C's, and an inversion center. Overall, a $\boldsymbol{C_{2h}}$ molecule.

4-2 **a.** Ethylene is a planar molecule, with C_2 axes through the C's, and perpendicular to the C−C bond both in the plane of the molecule and perpendicular to it. It also has a σ_h plane and two σ_d planes (arbitrarily assigned). Overall, a $\boldsymbol{D_{2h}}$ molecule.

b. Chloroethylene is also a planar molecule, with the only symmetry element the mirror plane of the molecule. Therefore, a $\boldsymbol{C_s}$ molecule.

c. 1,1-dichloroethylene has a C_2 axis coincident with the C−C bond, and two mirror planes, one the plane of the molecule and one perpendicular to the plane of the molecule through both C's. Overall, a $\boldsymbol{C_{2v}}$ molecule.

cis-1,2-dichloroethylene has a C_2 axis perpendicular to the C−C bond and in the plane of the molecule, two mirror planes (one the plane of the molecule and one perpendicular to the plane of the molecule and perpendicular to the C−C bond). Overall, a $\boldsymbol{C_{2v}}$ molecule.

trans-1,2-dichloroethylene has a C_2 axis perpendicular to the C−C bond and perpendicular to the plane of the molecule, a mirror plane in the plane of the molecule, and an inversion center. Overall, a $\boldsymbol{C_{2h}}$ molecule.

4-3 **a.** Acetylene has a C_∞ axis through all four atoms, an infinite number of perpendicular C_2 axes, a σ_h plane, and an infinite number of σ_d planes through all four atoms. Overall, a $\boldsymbol{D_{\infty h}}$ molecule.

b. Fluoroacetylene has only the C_∞ axis through all four atoms and an infinite number of mirror planes, also through all four atoms. Overall, a $\boldsymbol{C_{\infty v}}$ molecule.

c. Methylacetylene has a C_3 axis through the carbons and three σ_v planes, each including one hydrogen and all three C's. Overall, a $\boldsymbol{C_{3v}}$ molecule.

d. 3-Chloropropene (assuming a rigid molecule, no rotation around the C−C bond) has no rotation axes and only one mirror plane through Cl and all three C atoms. Overall, a C_s molecule.

e. Phenylacetylene (again assuming no internal rotation) has a C_2 axis down the long axis of the molecule and two mirror planes, one the plane of the benzene ring and the other perpendicular to it. C_{2v}.

4-4 **a.** Naphthalene has three perpendicular C_2 axes, and a horizontal mirror plane (regardless of which C_2 is taken as the principle axis), making it a D_{2h} molecule.

b. 1,8-dichloronaphthalene has only one C_2 axis, the C–C bond joining the two rings, and two mirror planes, making it a C_{2v} molecule.

c. 1,5-dichloronaphthalene has one C_2 axis perpendicular to the plane of the molecule, a horizontal mirror plane, and an inversion center; overall, C_{2h}.

d. 1,2-dichloronaphthalene has only the mirror plane of the molecule, and is a C_s molecule.

4-5 **a.** 1,1'-Dichloroferrocene has a C_2 axis parallel to the rings, perpendicular to the Cl–Fe–Cl σ_h mirror plane. It also has an inversion center. Overall, C_{2h}.

b. Dibenzenechromium has collinear C_6, C_3, and C_2 axes perpendicular to the rings, six perpendicular C_2 axes, and a σ_h plane, making it a D_{6h} molecule. It also has three σ_v and three σ_d planes, S_3 and S_6 axes, and an inversion center.

c. Benzenediphenylchromium has a mirror plane through the Cr and the diphenyl bridge bond and no other symmetry elements, so it is a C_s molecule.

d. H_3O^+ has the same symmetry as NH_3; a C_3 axis, and three σ_v planes for a C_{3v} molecule.

e. O_2F_2 has a C_2 axis perpendicular to the O–O bond and perpendicular to a line connecting the fluorines. With no other symmetry elements, it is a C_2 molecule.

f. Formaldehyde has a C_2 axis collinear with the C=O bond, a mirror plane including all the atoms and another perpendicular to the first and including the C and O atoms. Overall, C_{2v}.

g. S_8 has C_4 and C_2 axes perpendicular to the average plane of the ring, four C_2 axes through opposite bonds, and four mirror planes perpendicular to the ring, each including two S atoms. Overall, D_{4d}.

h. Borazine has a C_3 axis perpendicular to the plane of the ring, three perpendicular C_2 axes, and a horizontal mirror plane. Overall, D_{3h}.

i. Tris(oxalato)chromate(III) has a C_3 axis and three perpendicular C_2 axes, each splitting a C–C bond and passing through the Cr. Overall, $\boldsymbol{D_3}$.

j. A tennis ball has three perpendicular C_2 axes (one through the narrow portions of each segment, the others through the seams) and two mirror planes including the first rotation axis. Overall, $\boldsymbol{D_{2d}}$.

4-6 a. Cyclohexane in the chair conformation has a C_3 axis perpendicular to the average plane of the ring, three perpendicular C_2 axes between the carbons, and three σ_v planes, each including the C_3 axis and one of the C_2 axes. Overall, $\boldsymbol{D_{3d}}$.

b. Tetrachloroallene has three perpendicular C_2 axes, one collinear with the double bonds and the other two at 45° to the Cl–C–Cl planes. It also has two σ_v planes, one defined by each Cl–C–Cl group. Overall, $\boldsymbol{D_{2d}}$.

c. The sulfate ion is a tetrahedral shape. $\boldsymbol{T_d}$.

d. A snowflake has hexagonal symmetry, similar to a benzene ring, and has collinear C_6, C_3, and C_2 axes, six perpendicular C_2 axes, and a horizontal mirror plane. Overall, $\boldsymbol{D_{6h}}$.

e. Diborane has three perpendicular C_2 axes and three perpendicular mirror planes. $\boldsymbol{D_{2h}}$.

f. 1,3,5-tribromobenzene has a C_3 axis perpendicular to the plane of the ring, three perpendicular C_2 axes, and a horizontal mirror plane. $\mathbf{D_{3h}}$.

1,2,3-tribromobenzene has a C_2 axis through the center Cl and two perpendicular mirror planes which include this axis. $\boldsymbol{C_{2v}}$.

1,2,4-tribromobenzene has only the plane of the ring as a mirror plane. $\boldsymbol{C_s}$.

g. A tetrahedron inscribed in a cube has T_d symmetry (see Figure 4-6).

h. The left and right ends of B_3H_8 are staggered with respect to each other. There is a C_2 axis through the borons. In addition, there are two planes of symmetry, each containing four H atoms, and two C_2 axes between these planes and perpendicular to the original C_2. The point group is D_{2d}.

4-7 a. A sheet of typing paper has three perpendicular C_2 axes and three perpendicular mirror planes. Overall, $\boldsymbol{D_{2h}}$.

b. An Erlenmeyer flask has an infinite-fold rotation axis and an infinite number of σ_v planes, $\boldsymbol{C_{\infty v}}$.

c. A screw has no symmetry operations other than the identity, for a $\boldsymbol{C_1}$ classification.

d. The number 96 (with the correct type font) has a C_2 axis perpendicular to the plane of the paper and a mirror plane in the plane of the paper, making it $\boldsymbol{C_{2h}}$.

e. Your choice--the list is too long to attempt to answer it here.

f. A pair of eyeglasses has only a vertical mirror plane. $\boldsymbol{C_s}$.

g. A 5-pointed star has a C_5 axis, five perpendicular C_2 axes, one horizontal and five vertical mirror planes. $\boldsymbol{D_{5h}}$.

h. A fork has only a mirror plane perpendicular to the average plane of the fork. $\boldsymbol{C_s}$.

i. Captain Ahab has no symmetry operation other than the identity. $\boldsymbol{C_1}$.

j. A metal washer has a C_∞ axis, an infinite number of perpendicular C_2 axes, an infinite number of σ_v mirror planes, and a horizontal mirror plane. Overall, $\boldsymbol{D_{\infty h}}$.

4-8 **a.**

a. $VOCl_3$ has $\boldsymbol{C_{3v}}$ symmetry.

b. PCl_3 has $\boldsymbol{C_{3v}}$ symmetry.

c. SOF_4 has $\boldsymbol{C_{2v}}$ symmetry

d. ClO_2^- has $\boldsymbol{C_{2v}}$ symmetry.

e. ClO_3^- has $\boldsymbol{C_{3v}}$ symmetry.

f. P_4O_6 has $\boldsymbol{T_d}$ symmetry.

b.

a. PH_3 has $\boldsymbol{C_{3v}}$ symmetry.

b. H_2Se has $\boldsymbol{C_{2v}}$ symmetry.

c. SeF_4 has $\boldsymbol{C_{2v}}$ symmetry.

d. PF_5 has $\boldsymbol{D_{3h}}$ symmetry.

e. ICl_4^- has $\boldsymbol{D_{4h}}$ symmetry.

f. XeO_3 has $\boldsymbol{C_{3v}}$ symmetry.

g. NO_3^- has $\boldsymbol{D_{3h}}$ symmetry.

h. $SnCl_2$ has $\boldsymbol{C_{2v}}$ symmetry in the vapor phase.

i. PO_4^{3-} has $\boldsymbol{T_d}$ symmetry.

j. SF_6 has $\boldsymbol{O_h}$ symmetry.

k. IF_5 has C_{4v} symmetry.

l. ICl_3 has C_{2v} symmetry.

m. $S_2O_3^{2-}$ has C_{3v} symmetry.

n. BF_2Cl has C_{2v} symmetry.

4-9 **a.** CO_2 has $D_{\infty h}$ symmetry.

SO_3 has D_{3h} symmetry.

CH_4 has T_d symmetry.

PCl_5 has D_{3h} symmetry.

SF_6 has O_h symmetry.

IF_7 has D_{5h} symmetry.

TaF_8^{3-} has D_{4d} symmetry.

b. CO_2 has $D_{\infty h}$ symmetry.

COF_2 has C_{2v} symmetry.

NO_2^- has C_{2v} symmetry.

SO_3 has D_{3h} symmetry.

SNF_3 has C_{3v} symmetry.

SO_2Cl_2 has C_{2v} symmetry.

XeO_3 has C_{3v} symmetry.

SO_4^{2-} has T_d symmetry.

SOF_4 has C_{2v} symmetry.

ClO_2F_3 has C_{2v} symmetry.

XeO_3F_2 has D_{3h} symmetry.

IOF_5 has C_{4v} symmetry.

4-10 **a.** p_x has $C_{\infty v}$ symmetry. Ignoring the difference in sign between the two lobes, it is $D_{\infty h}$.

b. d_{xy} has D_{2h} symmetry. Ignoring the signs, it is D_{4h}.

c. $d_{x^2-y^2}$ has D_{2h} symmetry. Ignoring the signs, it is D_{4h}.

d. d_{z^2} has $D_{\infty h}$ symmetry.

4-11 The superimposed octahedron and cube show the matching symmetry elements. The descriptions below are for the elements of a cube; each element also applies to the octahedron.

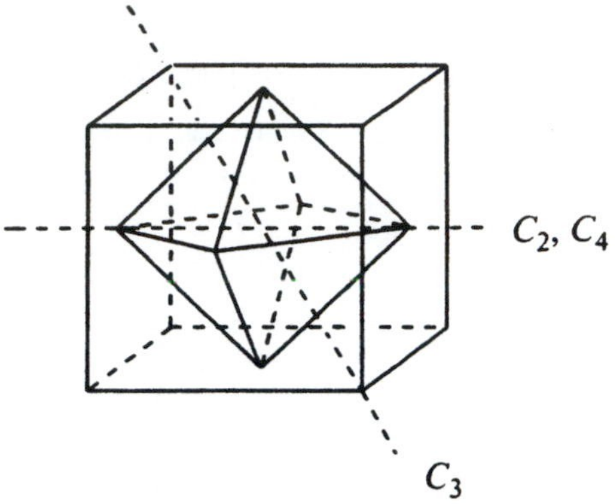

E	Every object has an identity operation.
$8C_3$	Diagonals through opposite corners of the cube are C_3 axes.
$6C_2$	Lines bisecting opposite edges are C_2 axes.
$6C_4$	Lines through the centers of opposite faces are C_4 axes. Although there are only three such lines, there are six axes, counting C_4^3.
$3C_2$	$(= C_4^2)$The lines through the centers of opposite faces are C_4 axes as well as C_2 axes.
i	The center of the cube is the inversion center.
$6S_4$	The C_4 axes are also S_4 axes.
$8S_6$	The C_3 axes are also S_6 axes.
$3\sigma_h$	These mirror planes are parallel to the faces of the cube.
$6\sigma_d$	These mirror planes are through two opposite edges.

4-12 **a.** C_{2h} molecules have E, C_2, i, and σ_h operations.

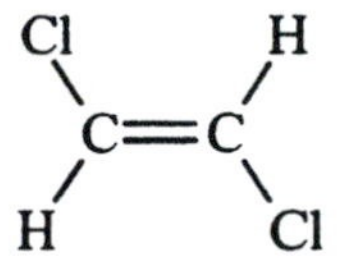

b. E: $\begin{matrix} 1 & 0 & 0 \\ 0 & 1 & 0 \\ 0 & 0 & 1 \end{matrix}$ C_2: $\begin{matrix} -1 & 0 & 0 \\ 0 & -1 & 0 \\ 0 & 0 & 1 \end{matrix}$ i: $\begin{matrix} -1 & 0 & 0 \\ 0 & -1 & 0 \\ 0 & 0 & -1 \end{matrix}$ σ_h: $\begin{matrix} 1 & 0 & 0 \\ 0 & 1 & 0 \\ 0 & 0 & -1 \end{matrix}$

c. These matrices can be block diagonalized into three 1 × 1 matrices, with the representations shown in the table.

	$\chi(E)$	$\chi(C_2)$	$\chi(i)$	$\chi(\sigma_h)$	
B_u	1	-1	-1	1	from the x and y coefficients
A_u	1	1	-1	-1	from the z coefficients

The total is $\Gamma = 2B_u + A_u$.

d. Multiplying B_u and A_u:
$1 \times 1 + (-1) \times 1 + (-1) \times (-1) + 1 \times (-1) = 0$, proving they are orthogonal.

4-13 **a.** D_{2h} molecules have E, $C_2(z)$, $C_2(y)$, $C_2(x)$, i, $\sigma(xy)$, $\sigma(xz)$, and $\sigma(yz)$ operations.

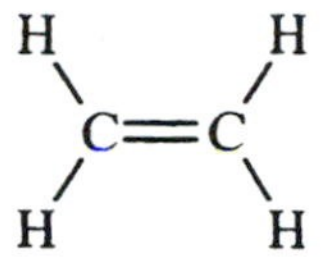

b.

E: $\begin{matrix} 1 & 0 & 0 \\ 0 & 1 & 0 \\ 0 & 0 & 1 \end{matrix}$ $C_2(z)$: $\begin{matrix} -1 & 0 & 0 \\ 0 & -1 & 0 \\ 0 & 0 & 1 \end{matrix}$ $C_2(y)$: $\begin{matrix} -1 & 0 & 0 \\ 0 & 1 & 0 \\ 0 & 0 & -1 \end{matrix}$

$C_2(x)$: $\begin{matrix} 1 & 0 & 0 \\ 0 & -1 & 0 \\ 0 & 0 & -1 \end{matrix}$ i: $\begin{matrix} -1 & 0 & 0 \\ 0 & -1 & 0 \\ 0 & 0 & -1 \end{matrix}$ $\sigma(xy)$: $\begin{matrix} 1 & 0 & 0 \\ 0 & 1 & 0 \\ 0 & 0 & -1 \end{matrix}$

$\sigma(xz)$: $\begin{matrix} 1 & 0 & 0 \\ 0 & -1 & 0 \\ 0 & 0 & 1 \end{matrix}$ $\sigma(yz)$: $\begin{matrix} -1 & 0 & 0 \\ 0 & 1 & 0 \\ 0 & 0 & 1 \end{matrix}$

c.

	E	$C_2(z)$	$C_2(y)$	$C_2(x)$	i	$\sigma(xy)$	$\sigma(xz)$	$\sigma(yz)$
χ	3	-1	-1	-1	-3	1	1	1

d.

Γ_1	1	-1	-1	1	-1	1	1	-1	matching B_{3u}
Γ_2	1	-1	1	-1	-1	1	-1	1	matching B_{2u}
Γ_3	1	1	-1	-1	-1	-1	1	1	matching B_{1u}

e. $\Gamma_1 \times \Gamma_2 = 1 \times 1 + (-1) \times (-1) + (-1) \times 1 + 1 \times (-1) + (-1) \times (-1) + 1 \times 1 + 1 \times (-1) + (-1) \times 1 = 0$

$\Gamma_1 \times \Gamma_3 = 1 \times 1 + (-1) \times 1 + (-1) \times (-1) + 1 \times (-1) + (-1) \times (-1) + 1 \times (-1) + 1 \times 1 + (-1) \times 1 = 0$

$\Gamma_2 \times \Gamma_3 = 1 \times 1 + (-1) \times 1 + 1 \times (-1) + (-1) \times (-1) + (-1) \times (-1) + 1 \times (-1) + (-1) \times 1 + 1 \times 1 = 0$

4-14 **a.** $h = 8$ (the total number of symmetry operations)

b. $A_1 \times E = 1 \times 2 + 2 \times 1 \times 0 + 1 \times (-2) + 2 \times 1 \times 0 + 2 \times 1 \times 0 = 0$

$A_2 \times E = 1 \times 2 + 2 \times 1 \times 0 + 1 \times (-2) + 2 \times (-1) \times 0 + 2 \times (-1) \times 0 = 0$

$B_1 \times E = 1 \times 2 + 2 \times (-1) \times 0 + 1 \times (-2) + 2 \times 1 \times 0 + 2 \times (-1) \times 0 = 0$

$B_2 \times E = 1 \times 2 + 2 \times (-1) \times 0 + 1 \times (-2) + 2 \times (-1) \times 0 + 2 \times 1 \times 0 = 0$

c. E: $4 + 2 \times 0 + 4 + 2 \times 0 + 2 \times 0 = 8$

A_1: $1 + 2 \times 1 + 1 + 2 \times 1 + 2 \times 1 = 8$

A_2: $1 + 2 \times 1 + 1 + 2 \times 1 + 2 \times 1 = 8$

B_1: $1 + 2 \times 1 + 1 + 2 \times 1 + 2 \times 1 = 8$

B_2: $1 + 2 \times 1 + 1 + 2 \times 1 + 2 \times 1 = 8$

d. $\Gamma_1 = 2A_1 + B_1 + B_2 + E$:

A_1: $1/8[1 \times 6 + 2 \times 1 \times 0 + 1 \times 2 + 2 \times 1 \times 2 + 2 \times 1 \times 2] = 2$

A_2: $1/8[1 \times 6 + 2 \times 1 \times 0 + 1 \times 2 + 2 \times (-1) \times 2 + 2 \times (-1) \times 2] = 0$

B_1: $1/8[1 \times 6 + 2 \times (-1) \times 0 + 1 \times 2 + 2 \times 1 \times 2 + 2 \times (-1) \times 2] = 1$

B_2: $1/8[1 \times 6 + 2 \times (-1) \times 0 + 1 \times 2 + 2 \times (-1) \times 2 + 2 \times 1 \times 2] = 1$

E: $1/8[2 \times 6 + 2 \times 0 \times 0 + (-2) \times 2 + 2 \times 0 \times 2 + 2 \times 0 \times 2] = 1$

$\Gamma_2 = 3A_1 + 2A_2 + B_1$:

A_1: $1/8[1 \times 6 + 2 \times 1 \times 4 + 1 \times 6 + 2 \times 1 \times 2 + 2 \times 1 \times 0] = 3$

A_2: $1/8[1 \times 6 + 2 \times 1 \times 4 + 1 \times 6 + 2 \times (-1) \times 2 + 2 \times (-1) \times 0] = 2$

B_1: $1/8[1 \times 6 + 2 \times (-1) \times 4 + 1 \times 6 + 2 \times 1 \times 2 + 2 \times (-1) \times 0] = 1$

B_2: $1/8[1 \times 6 + 2 \times (-1) \times 4 + 1 \times 6 + 2 \times (-1) \times 2 + 2 \times 1 \times 0] = 0$

E: $1/8[2 \times 6 + 2 \times 0 \times 4 + (-2) \times 6 + 2 \times 0 \times 2 + 2 \times 0 \times 0] = 0$

4-15 C_{3v}

$\Gamma_1 = 3A_1 + A_2 + E$:

A_1: $1/6[1 \times 6 + 2 \times 1 \times 3 + 3 \times 1 \times 2] = 3$

A_2: $1/6[1 \times 6 + 2 \times 1 \times 3 + 3 \times (-1) \times 2] = 1$

E: $1/6[2 \times 6 + 2 \times (-1) \times 3 + 3 \times 0 \times 2] = 1$

$\Gamma_2 = A_2 + 2E$:

A_1: $1/6[1 \times 5 + 2 \times 1 \times (-1) + 3 \times 1 \times (-1)] = 0$

A_2: $1/6[1 \times 5 + 2 \times 1 \times (-1) + 3 \times (-1) \times (-1)] = 1$

E: $1/6[2 \times 5 + 2 \times (-1) \times (-1) + 3 \times 0 \times (-1)] = 2$

O_h

$\Gamma_3 = A_{1g} + E_g + T_{1u}$:

A_{1g}: $1/48[6 + 0 + 0 + 12 + 6 + 0 + 0 + 0 + 12 + 12] = 1$

A_{2g}: $1/48[6 + 0 + 0 - 12 + 6 + 0 + 0 + 0 + 12 - 12] = 0$

E_g: $1/48[12 + 0 + 0 + 0 + 12 + 0 + 0 + 0 + 24 + 0] = 1$

T_{1g}: $1/48[18 + 0 + 0 + 12 - 6 + 0 + 0 + 0 - 12 - 12] = 0$

T_{2g}: $1/48[18 + 0 + 0 - 12 - 6 + 0 + 0 + 0 - 12 + 12] = 0$

A_{1u}: $1/48[6 + 0 + 0 + 12 + 6 + 0 + 0 + 0 - 12 - 12] = 0$

A_{2u}: $1/48[6 + 0 + 0 - 12 + 6 + 0 + 0 + 0 - 12 + 12] = 0$

E_u: $1/48[12 + 0 + 0 + 0 + 12 + 0 + 0 + 0 - 24 + 0] = 0$

T_{1u}: $1/48[18 + 0 + 0 + 12 - 6 + 0 + 0 + 0 + 12 + 12] = 1$

T_{2u}: $1/48[18 + 0 + 0 - 12 - 6 + 0 + 0 + 0 + 12 - 12] = 0$

4-16 The d_{xy} characters match the characters in the B_{2g} representation.

The $d_{x^2-y^2}$ characters match those of the B_{1g} representation

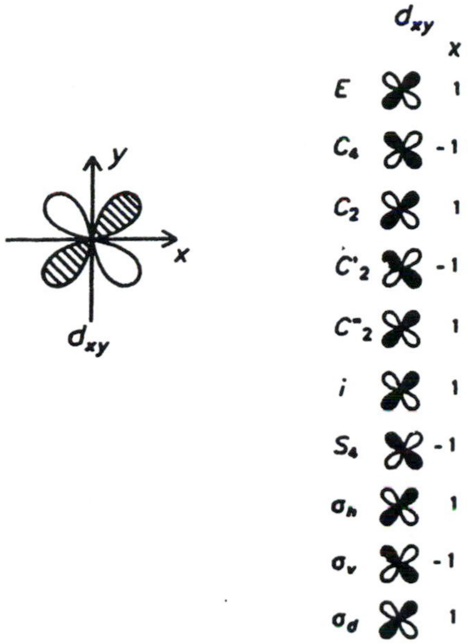

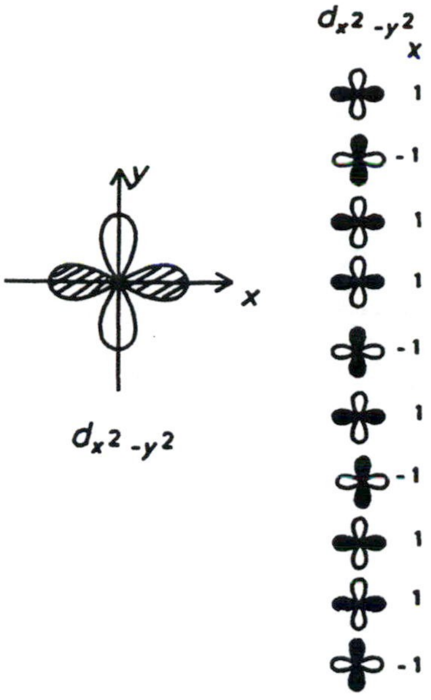

4-17 Chiral species from 4-5: **e** (O_2F_2) and **i** [tris(oxalato)chromate(III)]
From 4-6: none
From 4-7: **c** (screw) and **i** (Captain Ahab)
Other items from everyday life include a human foot (C_1), a piece of paper with two identical spiral designs printed on one side (C_2), and many organisms when small asymmetric details are considered.

4-18 **a.** *cis*-$Fe(CO)_4Cl_2$ has C_{2v} symmetry. The vectors for CO stretch have the representation Γ below:

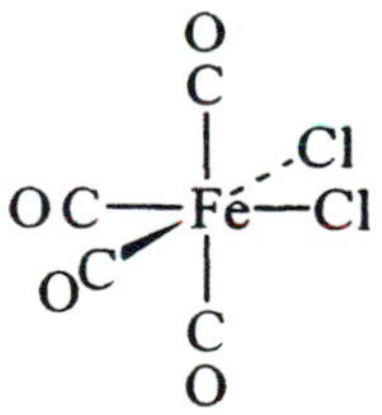

C_{2v}	E	C_2	$\sigma_v(xz)$	$\sigma_v'(yz)$	
Γ	4	0	2	2	
A_1	1	1	1	1	z
A_2	1	1	-1	-1	
B_1	1	-1	1	-1	x
B_2	1	-1	-1	1	y

$n(A_1) = 1/4\ [4 \times 1 + 0 \times 1 + 2 \times 1 + 2 \times 1] = 2$
$n(A_2) = 1/4\ [4 \times 1 + 0 \times 1 + 2 \times (-1) + 2 \times (-1)] = 0$
$n(B_1) = 1/4\ [4 \times 1 + 0 \times (-1) + 2 \times 1 + 2 \times (-1)] = 1$
$n(A_1) = 1/4\ [4 \times 1 + 0 \times (-1) + 2 \times (-1) + 2 \times 1] = 1$
$\Gamma = 2A_1 + B_1 + B_2$, all four IR active.

b. *trans*-$Fe(CO)_4Cl_2$ has D_{4h} symmetry.

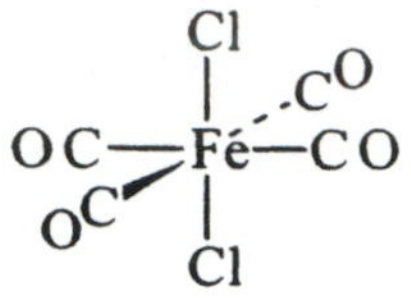

D_{4h}	E	$2C_4$	C_2	$2C_2'$	$2C_2''$	i	$2S_4$	σ_h	$2\sigma_v$	$2\sigma_d$	
Γ	4	0	0	2	0	0	0	4	2	0	
A_{2u}	1	1	1	-1	-1	-1	-1	-1	1	1	z
E_u	2	0	-2	0	0	-2	0	2	0	0	(x, y)

Omitting the operations that have zeroes in Γ:
$n(A_{2u}) = 1/16\ [4 \times 1 + 2 \times 2 \times (-1) + 4 \times (-1) + 2 \times 2 \times 1] = 0$
$n(E_u) = 1/16\ [4 \times 2 + 2 \times 2 \times 0 + 4 \times 2 + 2 \times 2 \times 0] = 1$ (IR active)

Note: In checking for IR-active bands, it is only necessary to check the irreducible representations having the same symmetry as x, y, or z, or a combination of them.

4-19 For SF_6, the axes of the sulfur should point at three of the fluorines. The fluorine axes can be chosen in any way, as long as one from each atom is directed toward the sulfur atom. Seven atoms, three axes each, for a total of 21.

O_h	E	$8C_3$	$6C_2$	$6C_4$	$3C_2$	i	$6S_4$	$8S_6$	$3\sigma_h$	$6\sigma_d$		
Γ	21	0	-1	3	-3	-3	-1	0	5	3		
T_{1u}	3	0	-1	1	-1	-3	-1	0	1	1	(x, y, z)	
T_{1g}	3	0	-1	1	-1	3	1	0	-1	-1	(R_x, R_y, R_z)	
A_{1g}	1	1	1	1	1	1	1	1	1	1		
A_{2u}	1	1	-1	-1	1	1	-1	1	1	-1		
E_g	2	-1	0	0	2	2	0	-1	2	0		$(2z^2 - x^2 - y^2, x^2 - y^2)$
T_{2u}	3	0	1	-1	-1	-3	1	0	1	-1		
T_{2g}	3	0	1	-1	-1	3	-1	0	-1	1		(xy, xz, yz)

Reduction of Γ gives $\Gamma = 3\ T_{1u} + T_{1g} + A_{1g} + E_g + T_{2g} + T_{2u}$. T_{1u} accounts for translation and also infrared active vibrational modes. T_{1g} is rotation. The remainder are vibrations that are infrared inactive.

4-20 **I** has C_2 symmetry, with a C_2 axis running right to left, perpendicular to the Cl, N, Cl, N and Cl, P, Cl, P faces.
II also has C_2 symmetry, with the same C_2 axis as **I.**
III has only an inversion center and C_i symmetry.

4-21 C_s symmetry requires one mirror plane. The only way to have this with five hydrogen atoms around a carbon atom is for three of the hydrogens to be in a trigonal plane with the other two on the opposite side of the carbon, as shown in the figures.

From the side: 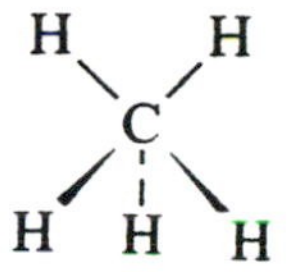

From the top: 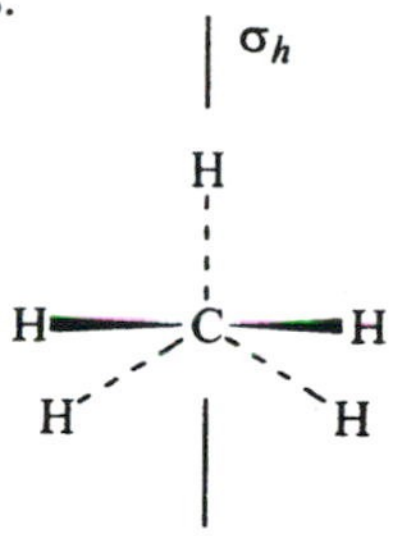

4-22 From the first diagram, this molecule appears to have S_6 symmetry. However, on closer inspection, all that the structure has is a center of inversion; the point group is C_i. (The reference calls this same point group S_2.)

4-23

Possible point groups	Example	
C_{3v}	CH_3F	12 examples
C_{2v}	CH_2F_2	6 examples
C_1	CHFClBr	1 example
C_s	CH_2FBr	12 examples
T_d	CF_4	4 examples

4-24 **a.** The S-C-C portion is linear, so the molecule has a C_3 axis along the line of these three atoms and three σ_v planes through these atoms and an F atom on each end, but no other symmetry elements. $\boldsymbol{C_{3v}}$

b. The molecule has only an inversion center, so it is $\boldsymbol{C_i}$. The inversion center is equivalent to an S_2 axis perpendicular to the average plane of the ring.

c. $M_2Cl_6Br_4$ also is $\boldsymbol{C_i}$.

d. This complex has a C_3 axis, splitting the three N atoms and the three P atoms (almost as drawn), but no other symmetry elements. $\boldsymbol{C_3}$

4-25 The Web of Science and SciFinder Scholar should be helpful, but simply searching for these symmetries also works. Some examples:

a. S_6: $Mo_2(SC_6H_2Me_3)_6$ (M. H. Chisholm, J. F. Corning, and J. C. Huffman, *J. Am. Chem. Soc.*, **1983**, *105*, 5924)
$Mo_2(NMe_2)_6$ (M. H. Chisholm, R. A. Cotton, B. A. Grenz, W. W. Reichert, L. W. Shive, and B. R. Stults, *J. Am. Chem. Soc.*, **1976**, *98*, 4469)
$[NaFe_6(OMe)_{12}(dbm)_6]^+$ (dbm = dibenzoylmethane, $C_6H_5COCCOC_6H_5$) (F. L. Abbati, A. Cornia, A. C. Fabretti, A. Caneschi, and D. Garreschi, *Inorg. Chem.*, **1998**, *37*, 1430)

b. T $Pt(CF_3)_4$, C_{44}

c. I_h C_{20}, C_{80}

Chapter 5
Molecular Orbitals

5-1 There are three possible bonding interactions:

p_z d_{z^2} p_y d_{yz} p_x d_{xz}

All other combinations have both positive and negative overlaps, resulting in no bonding.

5-2 O_2^{2-} has a single bond, with four electrons in the π* orbitals canceling those in the π orbitals.

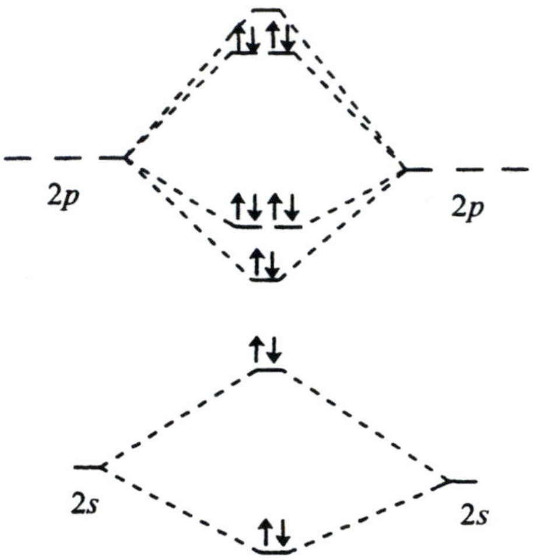

O_2^{2-}

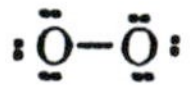

Lewis and Pauling

O_2^- has three electrons in the π* orbitals, and a bond order of 1.5. The Lewis structure has an unpaired electron and a bond order of 1.5. The Pauling structure has one 3-electron bond.

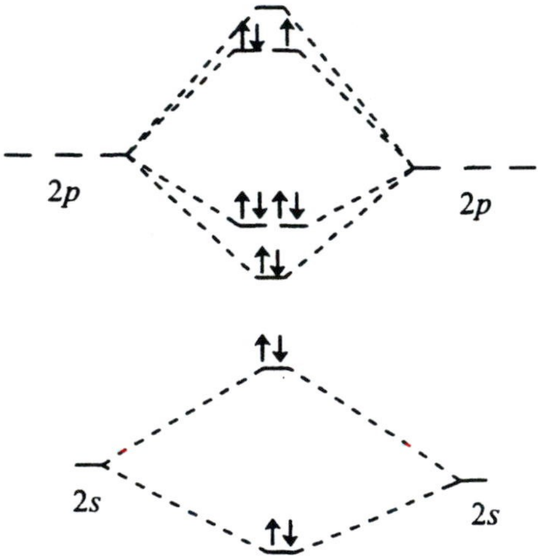

O_2^-

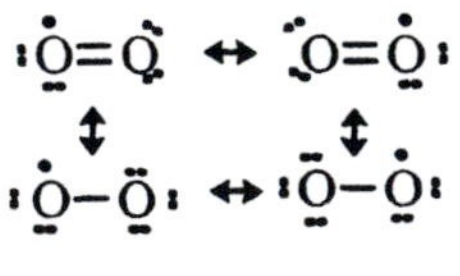

Lewis

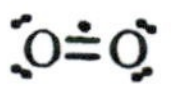

Pauling

O_2 has two unpaired electrons in the π^* orbitals, bond order of 2. The simple Lewis structure has all electrons paired, which does not match the experimental results. Two 3-electron bonds, in Pauling's version.

Bond lengths are therefore in the order $O_2^{2-} > O_2^- > O_2$, and bond strengths are the reverse of this order.

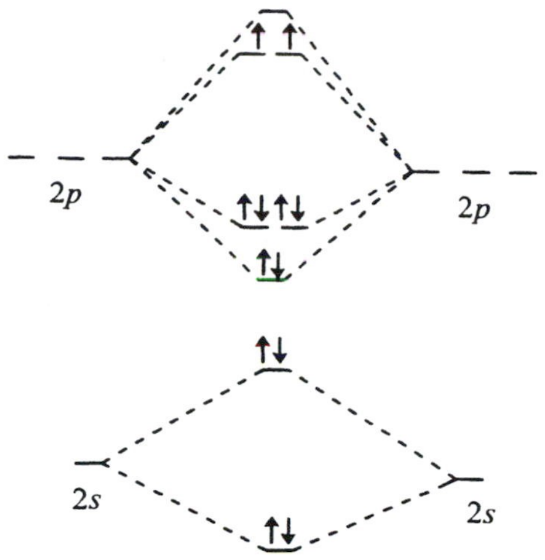

O_2

Lewis Pauling

5-3

	Bond order (from Figs. 5-5 and 5-7)	Bond distance (pm)	Unpaired electrons
C_2^{2-}	3	119	0
N_2^{2-}	2	122.4	2
O_2^{2-}	1	149 (very long)	0
O_2	2	120.7	2

The bond distance in N_2^{2-} is very close to the expected bond distance for a diatomic with 12 valence electrons, as shown in Figure 5-8.

5-4 **a.** The energy level diagram for NO is on the right. The odd electron is in a π_{2p}^* orbital.

b. O is more electronegative than N, so its orbitals are slightly lower in energy. The bonding orbitals are slightly more concentrated on O.

c. The bond order is 2.5, with one unpaired electron.

d. NO^+ Bond order = 3
shortest bond (106 *pm*)
NO Bond order = 2.5
intermediate (115 *pm*)

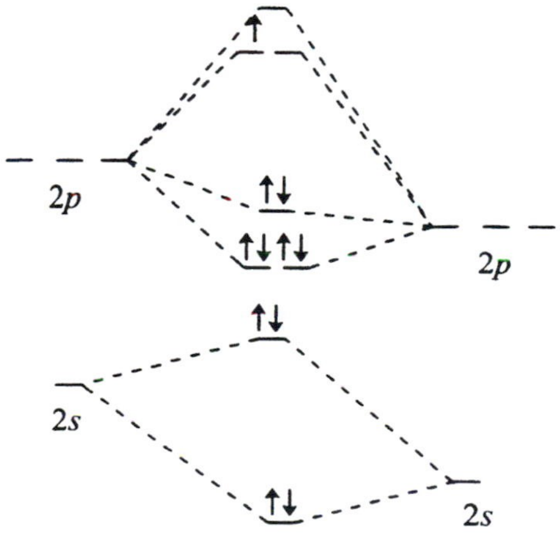

N NO O

NO^- Bond order = 2
longest bond, two electrons in antibonding orbitals, probably paramagnetic

5-5 **a.** The CN^- energy level diagram is similar to that of NO (Problem 5-4) without the antibonding electron.

b. The bond order is three, with no unpaired electrons.

c. The HOMO is the σ_{2p} orbital, which can interact with the 1*s* of the H^+, as in the diagram below. The bonding orbital has an energy near that of the π orbitals; the antibonding orbital becomes the highest energy orbital.

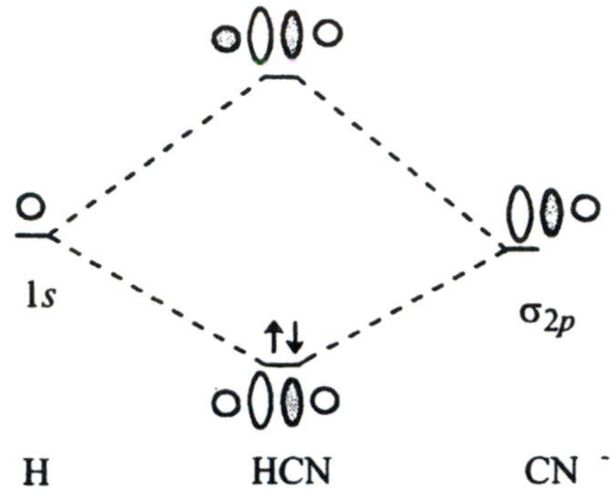

5-6 **a.** OF^- has 14 valence electrons, four in the π_{2p}* orbitals. (See the diagram in the answer to Problem 5-4.)

b. The net result is a single bond between two very electronegative atoms, and no unpaired electrons.

c. The concentration of electrons in the π* orbital is more on the O, so combination with the positive proton at that end is more likely. In fact, H^+ bonds to the oxygen atom, at an angle of 97°, as if the bonding were through a *p* orbital on O.

5-7 **a.** $KrBr^+$ energy level diagram is at the right.

b. The HOMO is polarized toward Br, since its energy is closer to that of the Br 4*p* orbital.

c. Bond order = 1

d. Kr is more electronegative. Its greater nuclear charge exerts a stronger pull on the shared electrons.

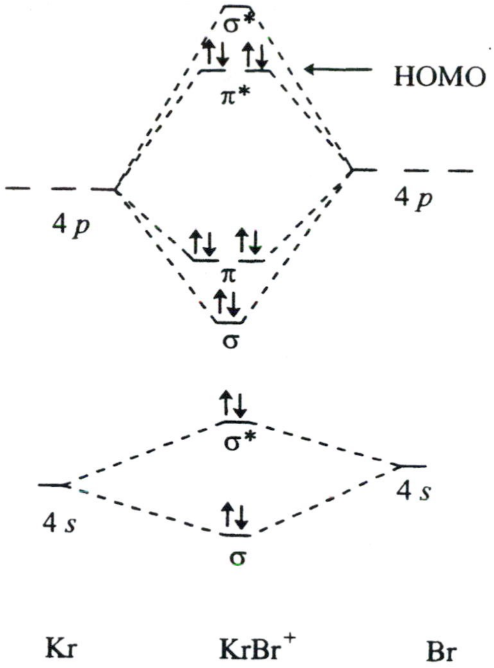

5-8 The molecular orbitals for SH^- are given below. Net, one bond.

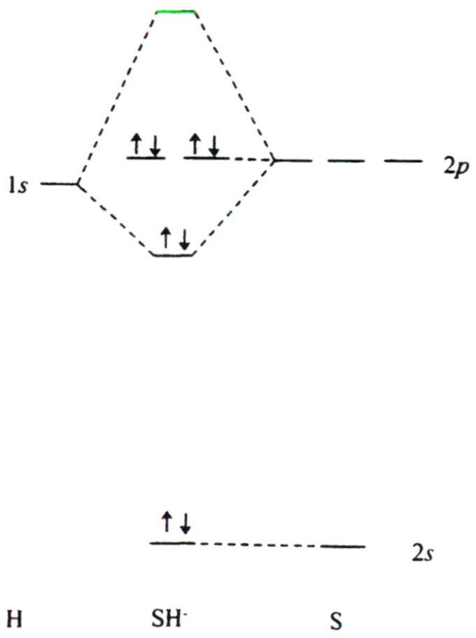

S orbital energies are − 12.0 (3*s*) and − 20.7 eV (3*p*), the 1*s* of H has an energy of − 13.6 eV.

5-9 **a.** The H—H group orbitals are 

that combines with the C 2*s*: 

and 

that combines with the C 2*p*: 

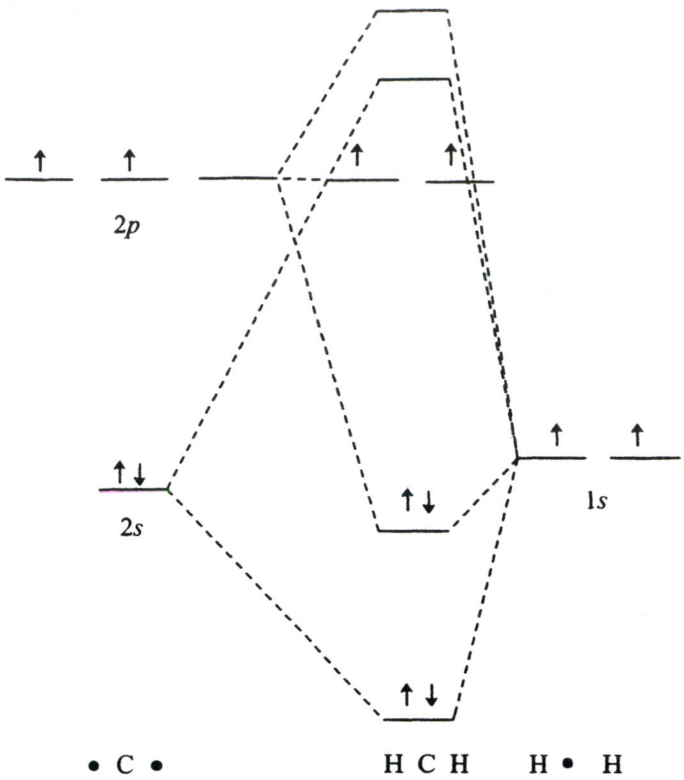

b. CH_2 is a paramagnetic diradical, with one electron in each of the p_x and p_y orbitals of the C. The singlet state, with these electrons paired in one of these nonbonding orbitals, has a significantly higher energy. Bonding orbitals are derived from the 2*s* of the C atom and the two 1*s* orbitals of the hydrogen atoms and from the 2*p* of the C atom and the two 1*s* orbitals from the hydrogen atoms. All these orbitals are close enough in energy (−10 to −19 eV) to interact easily. More careful calculations show that the H-C-H angle is about 130°.

5-10 BeF_2 uses *s* and *p* orbitals on all three atoms, and is isoelectronic with CO_2. The CO_2 orbitals in Figure 5-26 can be compared with these.

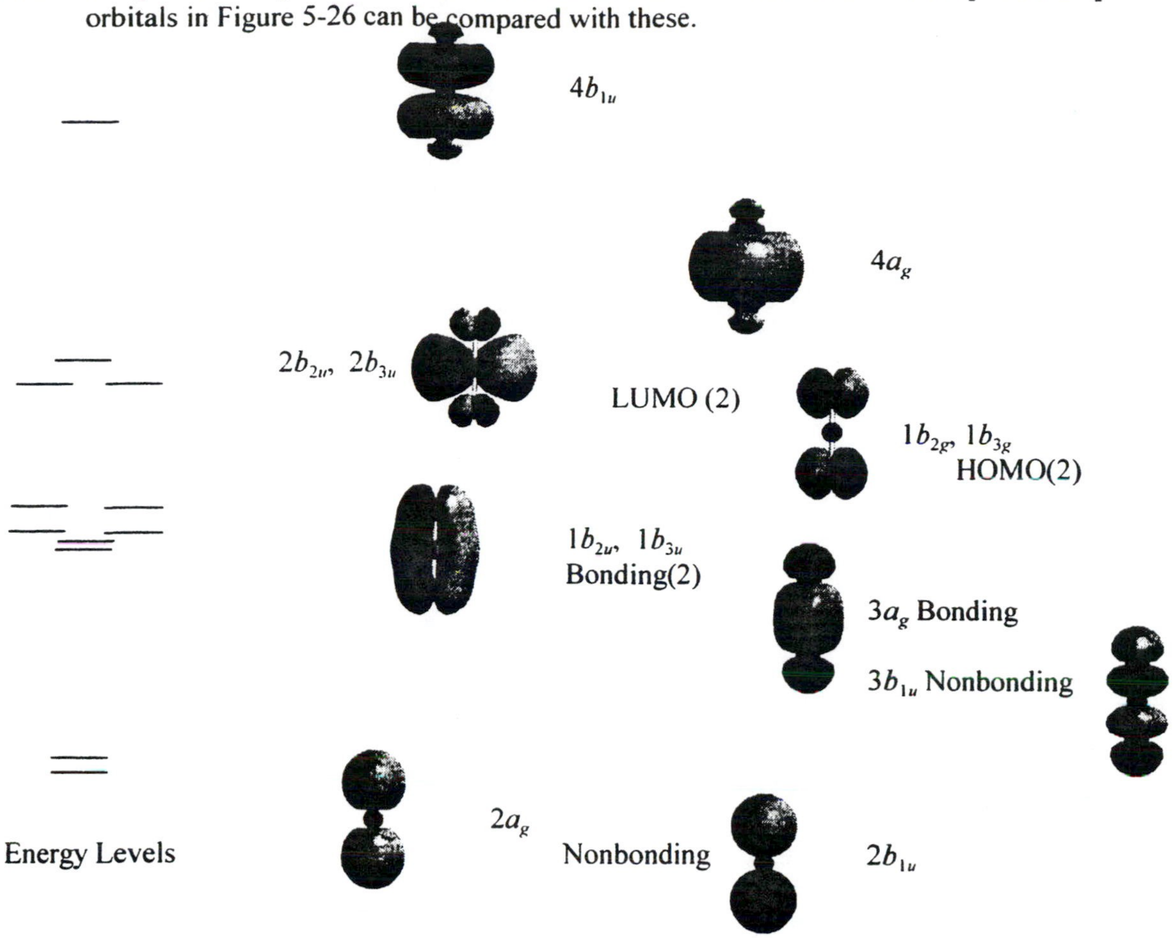

Of the occupied orbitals, there are three bonding (two π and one σ) and five nonbonding (two π and three σ). (References: W. R. Wadt and W. A. Goddard III, *J. Am. Chem.Soc.*, **1974,** *96*, 5996; R. Gleiter and R. Hoffmann, *J. Am. Chem. Soc.*, **1968**, *90*, 5457; C. W. Bauschlicher, Jr. and I. Shavitt, *J. Am. Chem. Soc.*, **1978**, *100*, 739.)

5-11 The group orbitals for the two F atoms are shown, each labeled with the Xe orbitals that match in symmetry.

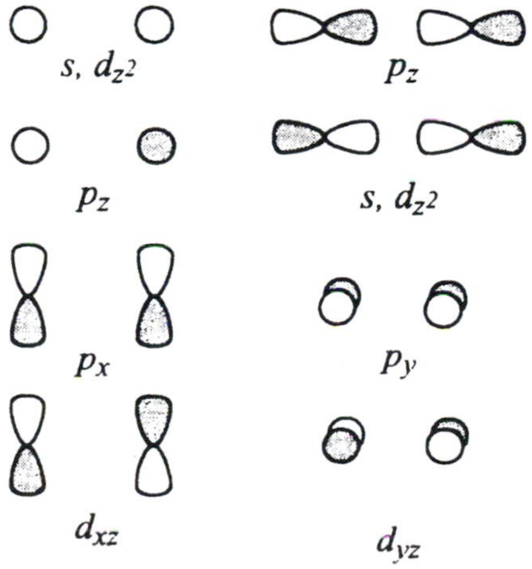

5-12 **a.** The energy level diagram for O_3 with the simple combinations of *s* and *p* orbitals is shown below.

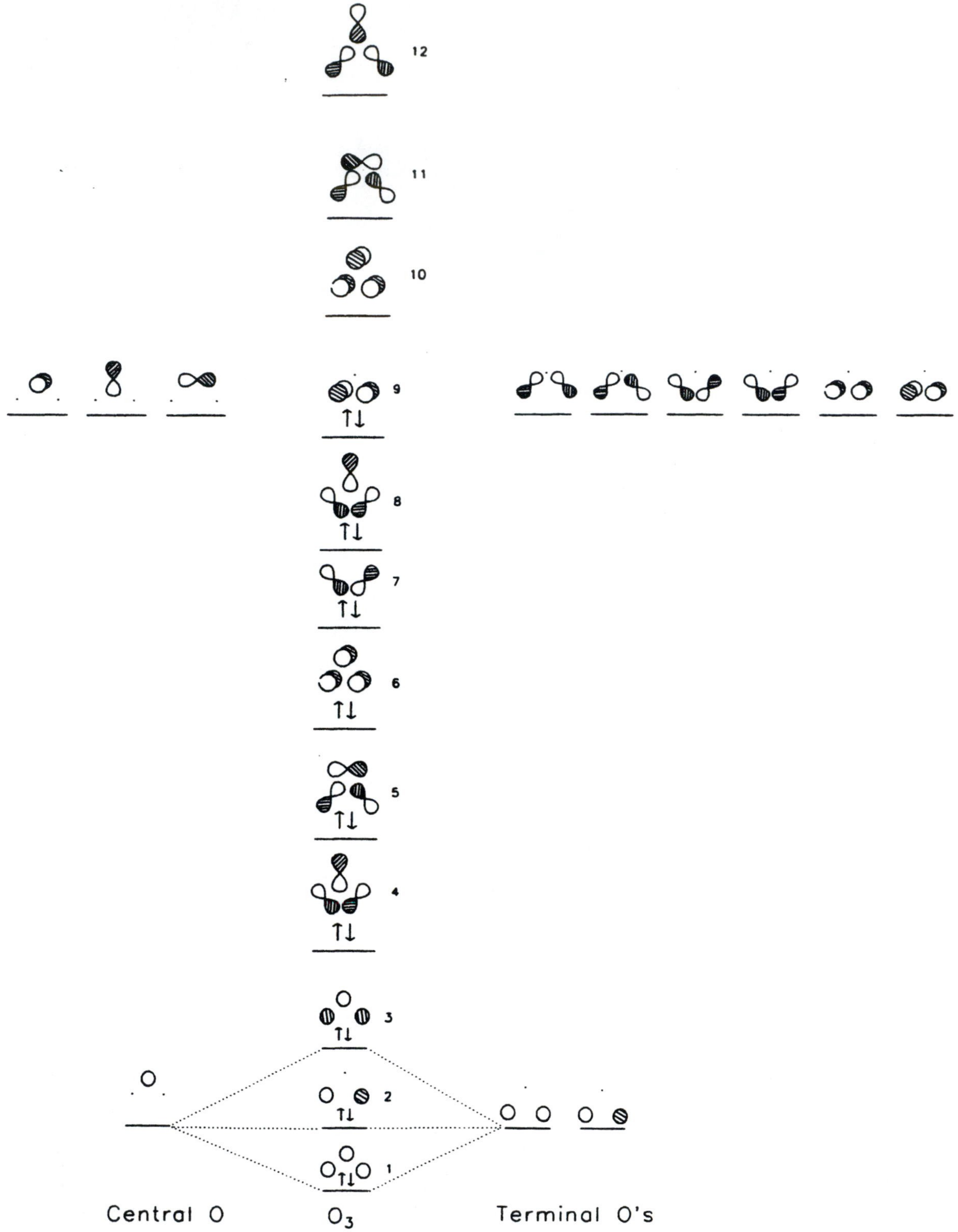

b. Mixing of *s* and *p* orbitals is fairly small, showing mostly in the four lowest orbitals. The order of orbitals may vary depending on the calculation method (for example, PM3 and AM1 methods reverse the orders of HOMO and HOMO-1).

5-13 As a cyclic (triangular) ion, H_3^+ has a pair of electrons in a bonding orbital and two vacant orbitals that are slightly antibonding:

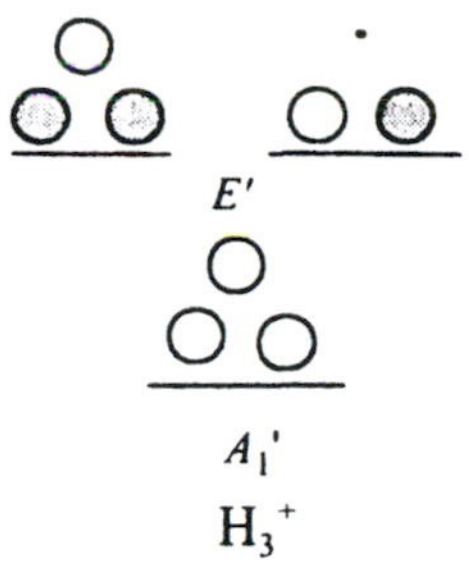

5-14 SO_3 has molecular orbitals similar to those of BF_3, described on pp. 154-156. The irreducible representations below are labeled for the oxygen orbitals.

D_{3h}	E	$2C_3$	$3C_2$	σ_h	$2S_3$	$3\sigma_v$	
$\Gamma(s)$	3	0	1	3	0	1	
$\Gamma(p_y)$	3	0	1	3	0	1	
$\Gamma(p_x)\parallel$	3	0	-1	3	0	-1	
$\Gamma(p_z)\perp$	3	0	-1	-3	0	1	
A_1'	1	1	1	1	1	1	
A_2'	1	1	-1	1	1	-1	
A_2''	1	1	-1	-1	-1	1	z
E'	2	-1	0	2	-1	0	(x, y)
E''	2	-1	0	-2	1	0	

σ $\Gamma(s) = A_1' + E'$ Sulfur s, p_x, and p_y

σ $\Gamma(p_y) = A_1' + E'$ Sulfur s, p_x, and p_y

$\pi_\parallel$ $\Gamma(p_x) = A_2' + E'$ Sulfur p_x and p_y

$\pi_\perp$ $\Gamma(p_z) = A_2'' + E''$ Sulfur p_z

5-15 The thiocyanate ion, SCN^-, has molecular orbitals similar to those of CO_2, but with more mixing between the *s* orbital of C and the *s* and *p* orbitals of S and N because the valence orbital potential energies of S are very close to those of C and those of N are only slightly lower. As a result, the net description of two double bonds developed in chapter 3, p. 54, fits the molecular orbital description as well. For cyanate, OCN^-, the *s* and *p* mix is similar on the N side, but less on the O side because the oxygen orbital energies are much lower. The structures described in chapter 3, p. 54, (a mix of two double bonds and O-C and C≡N) fit this ion also. For fulminate, CNO^-, the large difference between the C and O orbital energies makes mixing of the terminal orbitals difficult. As a result of this and the formal charge effects described in p. 55, the bonding in this ion is weak and it is stable only when combined in a complex with a metal ion.

5-16 The highest occupied orbitals in either SCN^- or OCN^- are π non-bonding orbitals (see Figure 5-26 for the similar CO_2 orbitals). Combination with H^+ or with metal ions depends on the energy match of these orbitals with those of the positive ion. The energy of the H orbital matches the energy of the N orbital better than either S or O orbitals. The S and N orbitals are close enough in energy to allow either end to bond with metal ions; slight differences dictate the preferred location. The S can also use the empty 3*d* orbitals to accept π bonding from some metal ions.

5-17 The CN^- molecular orbitals are similar to those of CO (Figure 5-14, p. 136 and problem 5-5), but with less difference between the C and N atomic orbital energies than between C and O orbitals. As a result, the HOMO should be more evenly balanced between the two atoms and bonding at both ends seems possible. The Prussian blue structures ($Fe_4[Fe(CN)_6]_3$ or $KFeFe(CN)_6$) have iron and CN^- in arrangements that have both Fe–C and Fe–N bonds, with uncertain Fe(II)-Fe(III) assignments.

5-18 **a.** The resonance structures were considered in Problem 3-3, showing bent structures with primarily double bond character in S=N and single bond in N–O or S–O. SNO^- is more stable on the basis of formal charges.

b. The molecular orbitals should be similar to those of O_3, with more mixing of *s* and *p* orbitals because of the difference between atomic orbital energies of S and O as terminal atoms. The π bonding, nonbonding, and antibonding orbitals are numbers 6, 9, and 10 in the ozone diagram in the answer to Problem 5-12.

π

S—N—O^-

π_n

S—N—O^-

π^*

S—N—O^-

π

N—S—O^-

π_n

N—S—O^-

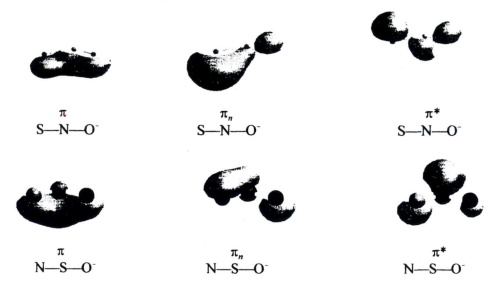

π^*

N—S—O^-

c. The calculated bond distances for these ions are

ion	N-S	S-O	N-O
SNO^-	146 *pm*		148 *pm*
NSO^-	171 *pm*	120 *pm*	

This indicates that SNO^- has nearly equal N-S and N-O bond lengths. Orbital energy arguments would favor a smaller S-N distance, which is the reverse of the formal charge arguments of problem 3-3.

5-19 SF_4 has C_{2v} symmetry. Treating the four F atoms as simple spherical orbitals, the reducible representation Γ_σ can be found and reduced to $\Gamma_\sigma = 2A_1 + B_1 + B_2$. Overall, the bonding orbitals can be dsp^2 or d^2sp, with the s and p_z or d_{z^2} orbitals with A_1 symmetry, p_x or d_{xy} with B_1 symmetry, and p_y or d_{yz} with B_2 symmetry. The p_z or d_{z^2} orbital remaining accounts for the lone pair. The use of the trigonal bipyramidal hybrids dsp^3 or d^3sp include the lone pair as one of the five locations.

C_{2v}	E	C_2	$\sigma(xz)$	$\sigma(yz)$	
Γ_σ	4	0	2	2	
A_1	1	1	1	1	z, z^2
A_2	1	1	-1	-1	
B_1	1	-1	1	-1	x, xz
B_2	1	-1	-1	1	y, yz

5-20 A square pyramidal molecule has the reducible representation $\Gamma = E + 2A_1 + B_1$.

C_{4v}	E	$2C_4$	C_2	$2\sigma_v$	$2\sigma_d$	
Γ	5	1	1	3	1	
E	2	0	-2	0	0	(x,y) (xz, yz)
A_1	1	1	1	1	1	z, z^2, x^2+y^2
B_1	1	-1	1	1	-1	x^2-y^2

There appear to be three possibilities for combining orbitals, depending on the details of their relative energies: dsp^3 (p_x and p_y for E, s and p_z for A_1, $d_{x^2-y^2}$ for B_1), d^2sp^2 (substituting d_{z^2} for p_z), and d^3sp (substituting d_{xz} and d_{yz} for p_x and p_y). Although d_{xz} and d_{yz} appear to work, they actually have their electron concentration between the B atoms, and therefore do not participate in σ bonding, so dsp^3 or d^2sp^2 fit better.

5-21 Square planar compounds have D_{4h} symmetry.

D_{4h}	E	$2C_4$	C_2	C_2'	C_2''	i	$2S_4$	σ_h	$2\sigma_v$	$2\sigma_d$	
Γ	4	0	0	2	0	0	0	4	2	0	
E_u	2	0	-2	0	0	-2	0	2	0	0	(x, y)
A_{1g}	1	1	1	1	1	1	1	1	1	1	z^2
B_{1g}	1	-1	1	1	-1	1	-1	1	1	-1	x^2-y^2

$$\Gamma = \underset{s,\, d_{z^2}}{\underset{\downarrow}{A_{1g}}} + \underset{d_{x^2-y^2}}{\underset{\downarrow}{B_{1g}}} + \underset{p_x,\, p_y}{\underset{\downarrow}{E_u}}$$

dsp^2 hybrids are the usual ones used for square planar compounds, although d^2p^2 is also possible. Since the d_{z^2} orbital does not extend far in the xy plane, it is less likely to participate in σ bonding.

5-22 **a.** PCl_5 has D_{3h} symmetry.

D_{3h}	E	$2C_3$	$3C_2$	σ_h	$2S_3$	$3\sigma_v$	
Γ	5	2	1	3	0	3	
E'	2	-1	0	2	-1	0	(x, y) $(x^2 - y^2, xy)$
A_1'	1	1	1	1	1	1	z^2
A_2''	1	1	-1	-1	-1	1	z

$\Gamma = E' + 2A_1' + A_2''$, so the hybrids are dsp^3 or d^3sp.

b. This could also be analyzed separately for the axial and the equatorial positions. The p_z and d_{z^2} orbitals can bond to the axial chlorines ($A_1' + A_2''$) and the s, p_x, and p_y orbitals or the s, $d_{x^2-y^2}$, and d_{xy} orbitals can bond to the equatorial chlorines (E').

c. The d_{z^2} orbital extends farther than the p orbitals, making the axial bonds a bit longer.

5-23 SO_3^{2-} is a trigonal pyramidal ion. The electron-dot pictures show two single bonds and one double bond between sulfur and oxygen, and ten electrons around the sulfur atom. The singly-bonded oxygen atoms have formal charges of 1−.

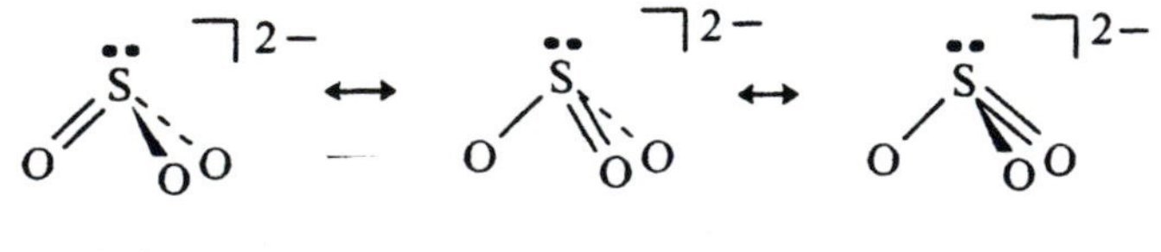

The molecular orbital picture is similar to that of ammonia, but the p orbitals of the oxygen atoms are added, with the double bond shared between all three oxygen atoms.

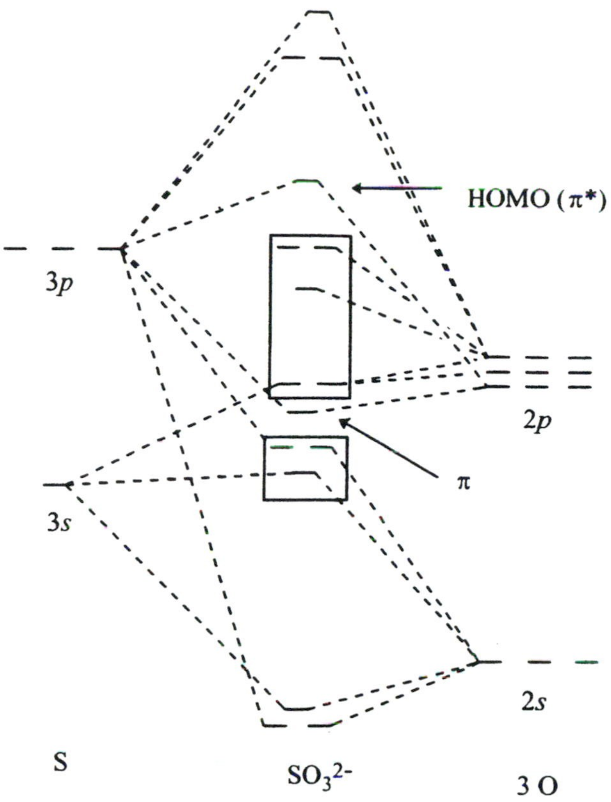

SO_3^{2-} is not described in the reference given, but SO_3 is. The addition of two electrons should lead to more "ionic" character in the sulfur-oxygen bonds, and structures like those below are likely. In all cases, the octet rule is not violated, but one oxygen atom is treated as if it were an oxide ion with eight electrons.

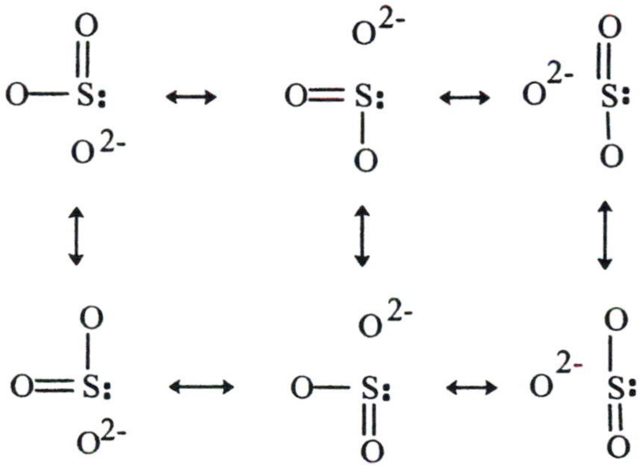

5-24 The full group theory treatment (D_{2h} symmetry), shown on pages 256-258, uses the two bridging hydrogens as one set for group orbitals and the four terminal hydrogens as another set. The bridging hydrogens have $\Gamma = A_g + B_{3u}$. The boron s orbitals have $\Gamma = A_g + B_{1u}$, the p_x orbitals (in the plane of the bridging hydrogens) have $\Gamma = B_{2g} + B_{3u}$, and the p_z orbitals (perpendicular to the plane of the bridging hydrogens) have $\Gamma = A_g + B_{1u}$. The boron A_g and

B_{3u} orbitals combine with the bridging hydrogen orbitals, resulting in two bonding and two antibonding orbitals. Two electrons in each of the bonding orbitals results in two bonds holding the molecule together through hydrogen bridges.

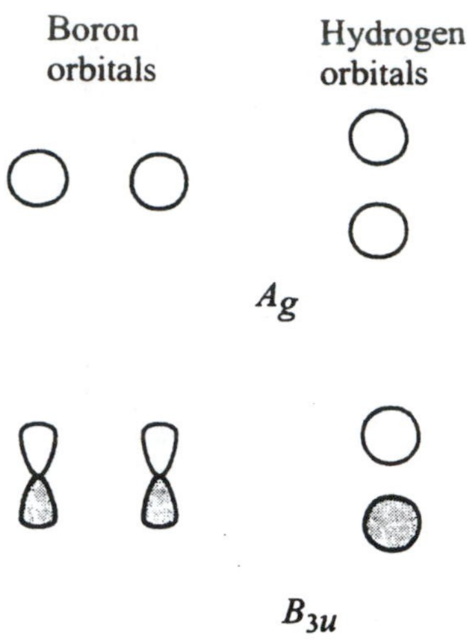

5-25 **a.** Cl_2^+ has one fewer electrons than Cl_2, so the π* levels have three, rather than four, electrons. As a result, Cl_2^+ has a bond order of 1.5, and the bond is shorter and stronger than that of Cl_2 (189 pm, compared with 199 pm for Cl_2).

b. Cl_4^+ has such an elongated rectangular shape (194 pm by 294 pm) that it must be essentially a Cl_2 and Cl_2^+ side by side, with only a weak attraction between them through the π* orbitals. The Cl-Cl bond in Cl_2 is 199 pm long; apparently the weak side-to-side bond draws off some of the antibonding electron density, strengthening and shortening the other two shorter Cl-Cl bonds.

5-26 **a.**

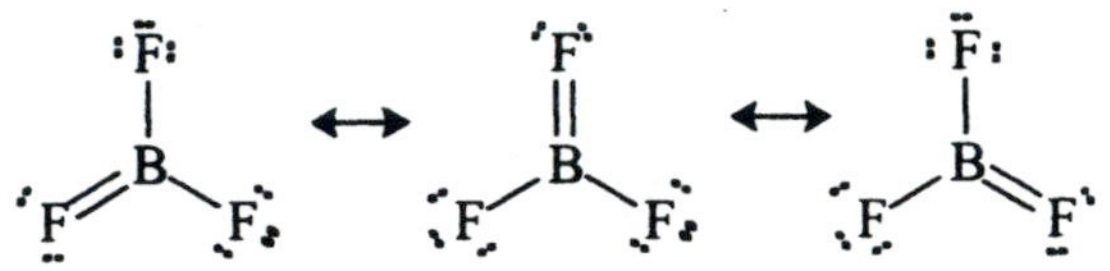

b. The $1a_2''$ orbital near the middle of the figure is the π-bonding orbital.

c. The LUMO, $2a_2''$, is the best orbital for accepting a lone pair.

d. The $1a_2''$ orbital is formed by adding all the p_z orbitals together. The $2a_2''$ orbital is formed by adding the B p_z orbital and subtracting the three F p_z orbitals.

5-27 **a.**

	ignoring orbital lobe signs	including orbital lobe signs
$1a_2''$	D_{3h}	C_{3v}
$2a_2''$	D_{3h}	C_{3v}
$1a_2'$	D_{3h}	C_{3h}
$1e''$	C_2	C_2

b. Results should be similar to Figure 5-33. The energies of some of the orbitals in the middle of the diagram are similar, and the order may vary with different calculation methods (for example, semi-empirical AM1 switches the order of $1e''$ and $2a_2'$).

c. The table of orbital contributions for each of the orbitals should show the same orbitals as in Figure 5-33. There may be some differences in contributions with different calculation methods, but they should be minor. Assignments to p_x, p_y, and p_z will also differ; the axes may be assigned differently by different programs.

Semi-empirical AM1 gives these as the major contributors to the specified orbitals:

	$3a_1'$	$4a_1'$	$1a_2''$	$1a_2'$	$2a_2''$
B	$2s$	$2s$	$2p_z$		$2p_z$
F	$2s$	$2s$, $2p_y$	$2p_z$	$2p_x$	$2p_z$

5-28 a. The energies and shapes of the orbitals depend on the method of calculation. CO in Figure 5-33 was calculated using the AM1 semi-empirical method, which shows the σ orbital as HOMO and has pear-shaped lobes in the π orbitals; the results below were obtained using STO-3G ab initio calculations, which reverse the order of the σ and π orbitals and gives π orbitals that are more nearly equal in size at the two ends. Energies are given in eV.

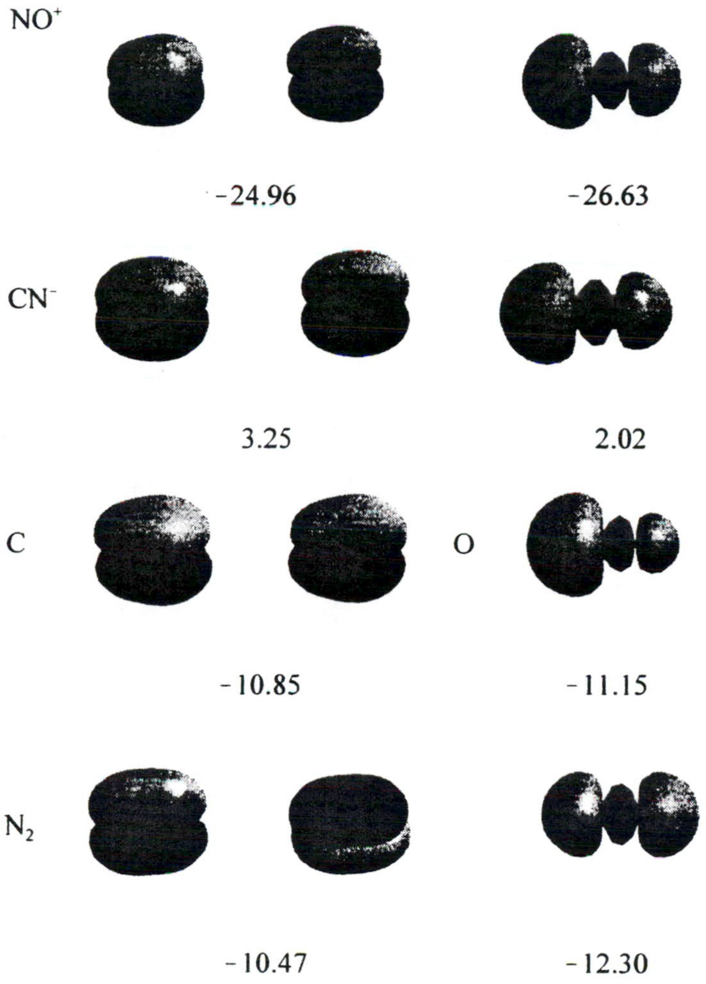

b. The STO-3G orbitals show only a slight concentration of electrons toward the more electronegative O in NO^+ and CO, and nearly equal distribution in CN^-. N_2, of course, has equal distribution.

c. The AM1 energies are listed below:

	σ^*	π^*	σ	π	σ^*	σ
		LUMO	HOMO			
CN^-	14.7	10.13	-3.13	-5.10	-9.37	-28.00
CO	5.28	0.94	-13.31	-16.30	-22.00	-41.2
N_2	6.03	1.00	-14.32	-16.19	-21.43	-41.39
NO^+	-4.42	-9.62	-26.13	-28.80	-35.80	-56.89

In this table, all the energies decrease as the atomic numbers increase, with CO and N_2 giving mixed results. There is considerable mixing of the σ orbitals, as shown in the order. Using the STO-3G calculations of part a, no mixing is apparent.

5-29 In one bonding orbital, the H *s* orbitals have the same sign and add to the Be *s* orbital in the HOMO-1 orbital. Subtracting the Be *s* orbital results in the antibonding LUMO. The difference between the two H *s* orbitals added to the Be p_z orbital results in the HOMO; Subtracting the Be p_z results in the LUMO+3 orbital. LUMO+1 and LUMO+2 are the Be p_x and p_y orbitals and are nonbonding in BeH_2. See the solution to Exercise 5-7 in Appendix A.

5-30 BeF_2 is similar to BeH_2, with the addition of π and π* orbitals from the p_x and p_y orbitals, extending over all three atoms. The F p_x orbitals with opposite signs do not combine with the Be orbitals, and neither do the p_y orbitals. They form the HOMO and HOMO+1 pair. The answer to Problem 5-10 shows more details.

5-31 The azide orbitals are very similar to the CO_2 orbitals, with some differences in contributions from the atomic orbitals because the CO_2 atomic orbitals do not have the identical energies of the nitrogen atoms. The two highest occupied orbitals of CO_2, BeF_2, and N_3^- all are p_x or p_y orbitals of the outer atoms with opposite signs, essentially nonbonding orbitals. The third orbital down has more *s* orbital contribution from the outer atoms than either of the other two; in those cases, the lower orbital energies of the atoms reduce that contribution. See also the solution to Exercise 5-6 in Appendix A.

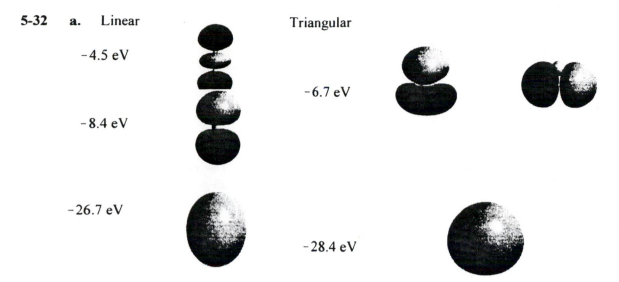

5-32 **a.** Linear

Triangular

−4.5 eV

−8.4 eV

−6.7 eV

−26.7 eV

−28.4 eV

b. Cyclic H_3^+ is slightly more stable than linear H_3^+, based on the energy of the lowest orbital in an AM1 calculation (−28.4 eV versus −26.7 eV).

Chapter 6
Acid-Base and Donor-Acceptor Chemistry

6-1

	Acid	Base	Definition
a.	BF_3	ClF	Lewis, solvent system
b.	$HClO_4$	CH_3CN	Lewis, Brønsted-Lowry
c.	ICl	PCl_5	Lewis, solvent system
d.	ClF_3	NOF	Lewis, solvent system
e.	SO_2	ClO_3^-	Lewis
f.	Pt	XeF_4	Lewis
g.	XeO_3	OH^-	Lewis
h.	SbF_5	HF	Lewis, solvent system
i.	Sn	NOCl	Lewis
j.	PtF_5	ClF_3	Lewis, solvent system
k.	CH_3COOH	$(benzyl)_3N$	Lewis, Brønsted-Lowry
l.	H_2O	BH_4^-	Lewis

6-2 Al^{3+} is acidic: $[Al(H_2O)_6]^{3+} + H_2O \rightleftharpoons [Al(H_2O)_5(OH)]^{2+} + H_3O^+$
The hydronium ions react with the basic bicarbonate to form CO_2:
$H_3O^+ + HCO_3^- \longrightarrow 2\ H_2O + CO_2\uparrow$ With pK_a values of 5.0 for $[Al(H_2O)_6]^{3+}$, 6.4 for H_2CO_3 and 2.0 for HSO_4^-, the pH is about 3, low enough to convert the bicarbonate to CO_2.

6-3 An increase in conductivity suggests that ions are formed:

$BrF_3 + AgF \rightleftharpoons BrF_4^- + Ag^+$

$BrF_3 + SnF_4 \rightleftharpoons BrF_2^+ + SnF_5^-$ or

$2\ BrF_3 + SnF_4 \rightleftharpoons 2\ BrF_2^+ + SnF_6^{2-}$

6-4 **a.** $3\ ICl \rightleftharpoons I_2Cl^+ + ICl_2^-$ (see Greenwood and Earnshaw, *Chemistry of the Elements*, 2nd ed., p. 827)

b. Both solutes increase the concentration of ions:

$NaCl + ICl \longrightarrow Na^+ + ICl_2^-$
(NaCl acts as a base)

$AlCl_3 + 2\ ICl \longrightarrow I_2Cl^+ + AlCl_4^-$ ($AlCl_3$ acts as an acid)

6-5 $SnCl_4 + 2\ Cl^- \longrightarrow SnCl_6^{2-}$ is the primary reaction. NH_4Cl in ICl forms NH_4^+ and ICl_2^-, and the chloride ions are then transferred to $SnCl_4$.

6-6 $KF + IF_5 \rightleftharpoons K^+ + IF_6^-$, and the ions conduct electricity.

6-7 $2\ H_2SO_4 \rightleftharpoons H_3SO_4^+ + HSO_4^-$ and $2\ H_3PO_4 \rightleftharpoons H_4PO_4^+ + H_2PO_4^-$ form enough ions to allow conductance in the pure acids.

6-8 **a.** The structure has the Br atoms in a staggered structure, resulting in an S_6 symmetry.

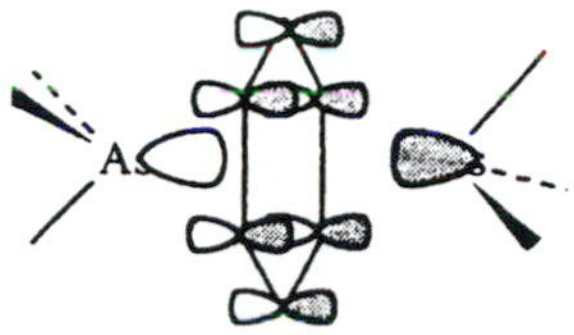

b. It may be easier to visualize this by using tetrahedral As. The fourth sp^3 orbitals point inward toward the benzene ring. If one is added and one is subtracted, they fit the symmetry of the π orbitals of the benzene ring to form bonding and antibonding orbitals. An essentially nonbonding orbital can be made from the two lobes with the same sign; one side has a bonding interaction with the benzene orbitals and the other has an antibonding interaction.

6-9 The very high electronegativity of O in comparison with Al pulls the bonding pair very close to O. This increases the repulsion between the bonding pairs and causes the large angle.

6-10 **a.** The methyl groups in $(CH_3)_3N—SO_3$ donate electrons to the nitrogen, making $(CH_3)_3N$ a stronger Lewis base and strengthening and shortening the N—S bond. the greater concentration of electrons in the N—S bond of $(CH_3)_3N—SO_3$ increases electron-electron (*bp-bp*) repulsions, opening up the N—S—O bond in comparison with $H_3N—SO_3$.

	$(CH_3)_3N—SO_3$	$H_3N—SO_3$
N—S	191.2 pm	195.7 pm
N—S—O	100.1°	97.6°

6-11 NO^- is isoelectronic with O_2 and has the electronic structure $\sigma^2\ \pi^2\ \pi^2\ \pi^{*1}\ \pi^{*1}$. Bonding with H^+ depends on which end of the π* orbital carries more electron density. Calculation shows slightly more electron density on N, making HNO the more likely (bent) molecule. NO^- readily dimerizes to $N_2O_2^{2-}$, with a *trans* configuration and a combination of π* orbitals from each of the monomers.

6-12 **a.** This is similar to the effects described in Section 6-2-8 for I_2. Br_2 forms charge-transfer complexes with donor solvents such as methanol.

b. The 500 nm band ($\pi^* \rightarrow \sigma^*$) should shift to shorter wave length (higher energy) because the difference in energy between the π^* and σ^* orbitals is greater in $Br_2 \bullet CH_3OH$ than in Br_2.

6-13. $AlF_3 + F^- \longrightarrow AlF_4^-$. The Na^+, AlF_4^- salt can dissolve in HF. When BF_3 is added, it has a stronger attraction for the fluoride ions, with the reaction $AlF_4^- + BF_3 \longrightarrow AlF_3\downarrow + BF_4^-$

6-14 Soft metal ions do not combine with oxygen as strongly as hard metal ions, so reactions like

$$2\ HgO \xrightarrow{\Delta} 2\ Hg + O_2\uparrow \qquad 2\ CuO \xrightarrow{\Delta} 2\ Cu + O_2\uparrow \qquad 2\ Ag_2O \xrightarrow{\Delta} 4\ Ag + O_2\uparrow$$

$$\text{and } CuO + C \xrightarrow{\Delta} Cu + CO\uparrow \qquad 2CuO + C \xrightarrow{\Delta} 2Cu + CO_2\uparrow$$

are more easily carried out. Reduction of some of the softer metals can be carried out with relatively low temperatures (campfires); some think use of rocks containing ores could have led to accidental reduction to the metals and discovery of the smelting process. Harder metals such as iron require much higher temperatures for the reduction process.

6-15 Hg is a much softer metal, and combines with the soft sulfide ion more strongly. Zinc and cadmium are more borderline metals, and combine with all the anions with more nearly equal facility.

6-16 When any of these salts vaporize, the vapor phase consists of molecules. When they are in the solid state, they are ionic, with some covalent properties. The liquid state is between the two, and can be made up of either ions, covalent molecules, or something between these two extremes. If the liquid is molecular, vaporization should be easier (molecules in the liquid phase being converted to molecules in the vapor phase). If the liquid is mostly ionic, vaporization requires greater molecular change and should be more difficult. Using this criterion, the most ionic liquids should be ZnF_2 and CdF_2 and the most molecular liquids should be HgF_2 and $HgCl_2$. On a more general view, mercury as the softest metal in the series forms the more molecular compounds and zinc as the hardest forms the more ionic compounds. Cadmium forms the most molecular compound with the borderline bromide.

6-17. a. pyridine + BF_3:

$$\Delta H = -(E_{py}E_{BF_3} + C_{py}C_{BF_3})$$
$$= -[(1.17)(9.88) + (6.40)(1.62)] = -21.9 \text{ kcal/mol or } -91.6 \text{ kJ/mol, about 10\% low}$$

pyridine + $B(CH_3)_3$:

$$\Delta H = -(E_{py}E_{B(CH_3)_3} + C_{py}C_{B(CH_3)_3})$$
$$= -[(1.17)(6.14) + (6.40)(1.70)] = -18.1 \text{ kcal/mol or } -75.7 \text{ kJ/mol, 5 to 18\% high}$$

b. Fluorine is electron-withdrawing, CH_3 electron-releasing. Therefore, B in BF_3 carries a greater positive charge and interacts more strongly with Lewis bases such as pyridine.

c. The harder acid BF_3 interacts more strongly with the borderline base pyridine.

6-18. $NH_3 + BF_3$:

$$\Delta H = -(E_{NH_3}E_{BF_3} + C_{NH_3}C_{BF_3})$$
$$= -[(1.36)(9.88) + (3.46)(1.62)] = -19.0 \text{ kcal/mol or } -79.5 \text{ kJ/mol}$$

$NH_3 + B(CH_3)_3$

$$\Delta H = -(E_{NH_3}E_{B(CH_3)_3} + C_{NH_3}C_{B(CH_3)_3})$$
$$= -[(1.36)(6.14) + (3.46)(1.70)] = -14.2 \text{ kcal/mol or } -59.4 \text{ kJ/mol}$$

The order is pyridine + BF_3 > NH_3 + BF_3 > pyridine + $B(CH_3)_3$ > NH_3 + $B(CH_3)_3$. The change from BF_3 to $B(CH_3)_3$ is larger than the change from pyridine to NH_3.

6-19 Absolute hardness parameters:

	I	A	η
BF_3	15.81	-3.5	9.7
NH_3	10.7	-5.6	8.2
C_5H_5N	9.3	-0.6	5.0
$N(CH_3)_3$	7.8	-4.8	6.3

The difference between the HOMO of pyridine (9.3) and the LUMO of BF_3 (-3.5) is smaller than the other possible interaction, so this pair has the largest $-\Delta H$, in spite of the difference in hardness. By comparison with the nitrogen compounds, $B(CH_3)_3$ is expected to have an absolute hardness lower than that of BF_3, approximately 7.5-8.

6-20 CsI and LiF fit Basolo's principle that ions of similar size and equal (but opposite) charge form the least soluble salts. CsF and LiI have ionic sizes that are very different, and they do not fit as well into an ionic lattice. In addition, CsI and LiF are soft-soft and hard-hard combinations, which combine better than the hard-soft LiI and soft-hard CsF.

6-21

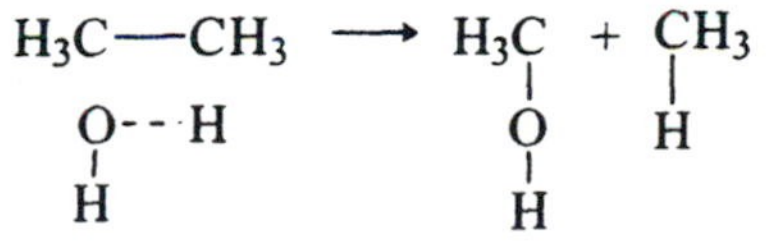

is unlikely. C is soft, O is hard.

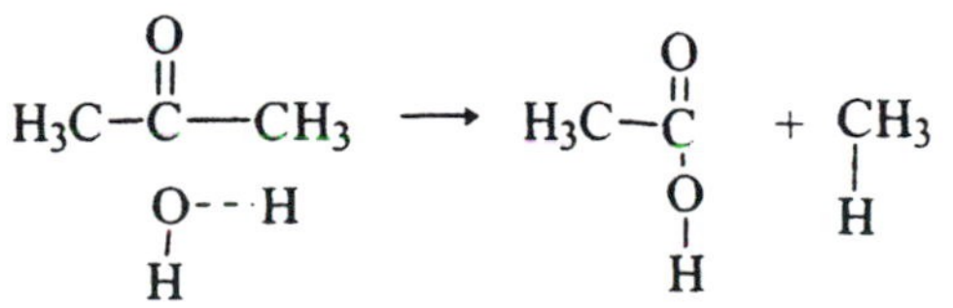

is more likely. Adding the carbonyl oxygen makes C harder, and C in CH_3 and the H atom are soft.

6-22 **a.** Solubilities: $MgSO_4 > CaSO_4 > SrSO_4 > BaSO_4$
Electrostatic forces predict the reverse order due to cation sizes, but the larger cations fit better with the large sulfate anion in the crystals. Hydration of the cations is strongest for Mg^{2+}, weakest for Ba^{2+}, agreeing with the solubility order.

b. Solubilities: $PbCl_2 > PbBr_2 > PbI_2 > PbS$
As a moderately soft cation, Pb^{2+} has stronger interactions with the softer anions (order $Cl^- < Br^- < I^- < S^{2-}$). In addition, hydration of the anions is largest for chloride, smallest for sulfide, based on size.

6-23 **a.** TeH_2 is the strongest acid, because Te is more electronegative than either Sn or Sb. Therefore, the hydrogen is more positive and acidic.

b. NH_3 is the strongest base because N is more electronegative than either P or Sb. The charge density is large enough that it not only holds its own hydrogens, but attracts another.

c. $(CH_3)_3N$ is the strongest base in the gas phase because the methyl groups contribute electron density to the nitrogen. In solution, the order is scrambled, probably due to solvation (pp. 200-201).

d. 4-Mepy > py > 2-Mepy Again, the methyl group adds electron density to the N. However, with methyl in the 2 position, steric hindrance makes bonding to BMe_3 more difficult.

6-24 In general, oxide ion reacts with water to form hydroxide: $O^{2-} + H_2O \longrightarrow 2\ OH^-$ unless other factors prevent it. In B_2O_3, the small, hard B^{3+} holds on to the oxide ions strongly. As a result, $B_2O_3 + 3\ H_2O \longrightarrow 2\ B(OH)_3 \rightleftharpoons H^+ + H_2BO_3^-$, and the solution is very weakly acidic ($pK_a = 9.25$). In Al_2O_3, the Al^{3+} ion is larger and softer. It can form either $[Al(OH)_4]^-$ (acting as an acid) or $[Al(H_2O)_6]^{3+}$ (acting as a base), depending on the other species in solution. Sc^{3+} is still larger and softer, so it combines better with water than with hydroxide ion. As a result, $Sc_2O_3 + 15\ H_2O \longrightarrow 2\ [Sc(H_2O)_6]^{3+} + 6\ OH^-$ is possible.

6-25 **a.** $CaH_2 + 2\ H_2O \longrightarrow Ca^{2+} + 2\ H_2\uparrow + 2\ OH^-$ Calcium has a lower electronegativity than hydrogen, so CaH_2 is $Ca^{2+}(H^-)_2$ and the hydride ions react with the positive hydrogens of water.

b. $HBr + H_2O \longrightarrow H_3O^+ + Br^-$ Bromine has a larger electro-negativity than hydrogen, so the hydrogen is strongly positive and is readily lost to the lone pair of water.

c. $H_2S + H_2O \rightleftharpoons H_3O^+ + HS^-$ Sulfur is slightly more electronegative than hydrogen, and the positive hydrogen can dissociate to a small extent.

d. $CH_4 + H_2O \longrightarrow$ no reaction. The C-H bond is almost nonpolar; the hydrogens are not positive enough to be attracted to the water lone pair.

6-26 $BF_3 > B(CH_3)_3 > B(C_2H_5)_3 > B(C_6H_2(CH_3)_3)_3$ Alkyl groups are electron-donating and increase the electron density on B and reduce the attraction for the lone pair of NH_3. The large mesitylene groups reduce the adduct formation because they are too bulky to fold back readily into the required tetrahedral geometry.

6-27. **a.** CH_3NH_2 is a stronger base. The methyl group pushes electron density onto the nitrogen.

b. Although 2-methylpyridine is the stronger base with smaller acid molecules, the methyl group interferes with adduct formation with trimethylboron (F-strain) and the pyridine-trimethylboron formation is stronger.

c. Trimethylboron forms a stronger adduct with ammonia because the three phenyl rings of triphenylboron cannot bend back readily to allow the boron to become tetrahedral (B-strain).

6-28. **a.** With the acids listed in order of increasing acidity:

	H_3AsO_4	H_2SO_3	H_2SO_4	$HMnO_4$
pK_a (9-7*n*)	2	2	-5	-12
pK_a (8-5*n*)	3	3	-2	-7
pK_a (exptl)	9.2	2.2	1.8	-11

b. With the acids listed in order of increasing acidity:

	HClO	$HClO_2$	$HClO_3$	$HClO_4$
pK_a (9-7*n*)	9	2	-5	-12
pK_a (8-5*n*)	8	3	-2	-7
pK_a (exptl)	7.4	2	-1	-10

6-29 Dimethylamine acts as a weak base in water, with a very small amount of OH^- provided by the reaction $(CH_3)_2NH + H_2O \rightleftharpoons (CH_3)_2NH_2^+ + OH^-$. Acetic acid is a stronger acid than water, so dimethylamine acts as a stronger base and the reaction $(CH_3)_2NH + HOAc \longrightarrow (CH_3)_2NH_2^+ + OAc^-$ goes to completion. 2-Butanone is a neutral solvent; there is no significant acid-base reaction with dimethylamine.

6-30. SbF_5 in HF reacts to increase the H^+ concentration and decrease H_0:

$SbF_5 + HF \rightleftharpoons H^+ + SbF_6^-$ These ions then can react with alkenes.

6-31 **a.** The energy diagram and the orbitals are shown below.

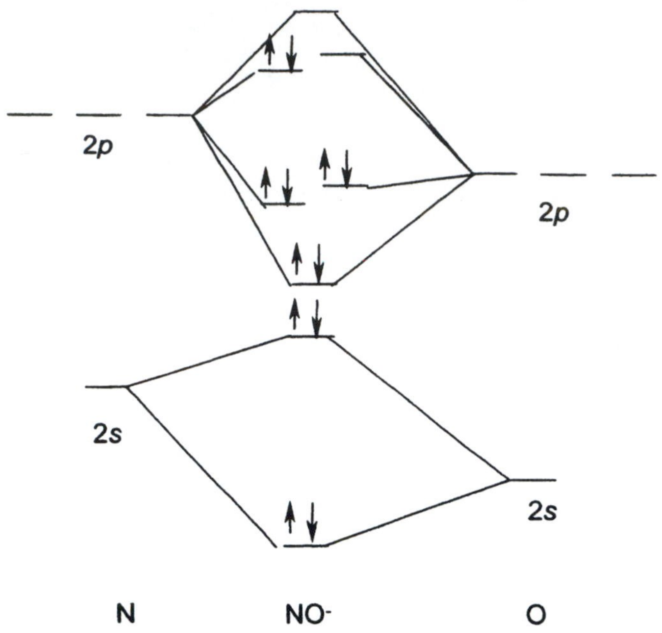

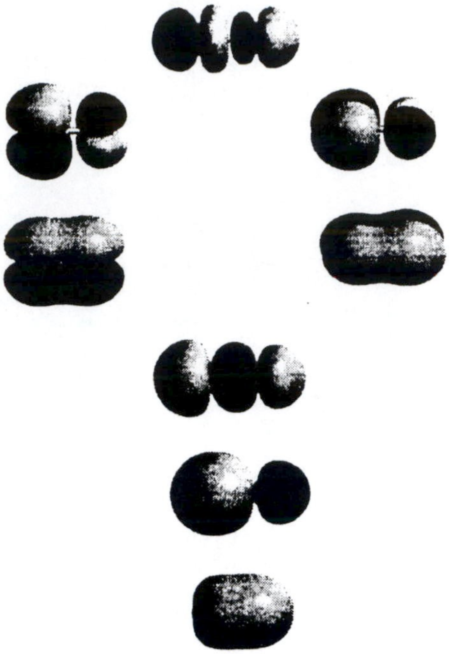

Simple calculation of the singlet state (shown) and the triplet state (two unpaired electrons in the π^* orbitals) show the triplet as slightly more stable. In either case, the HOMO is the π^*, with larger concentration of electron density on N.

b. The π^* HOMO can combine with the empty H^+ $1s$ orbital, forming HNO, a bent molecule. Comparing HNO and NOH in a PM3 calculation:

HNO NOH

HOMO

HOMO-1

58 kJ/mol 205 kJ/mol Enthalpy of formation

Neither is stable, but HNO is more stable than NOH

6-32

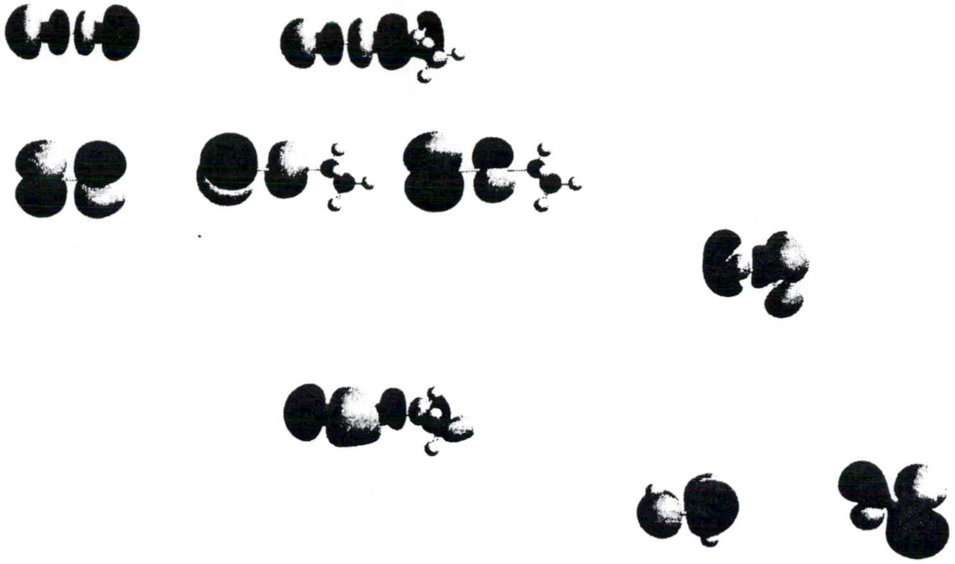

The interaction of the HOMO of the methanol and the LUMO of the Br_2 results in the LUMO and HOMO−2 orbitals of the adduct (the HOMO and HOMO−1 of the Br_2 form the intermediate orbitals shown). The geometry shown has the Br_2 at approximately a trigonal angle (H-O-Br is 113° and C-O-Br is 106° in this PM3 calculation).

Chapter 7
The Crystalline Solid State

7-1 Solutions of alkali halides in water conduct electricity. This does not prove that they are ionic as solids, but is suggestive of ions in the solid state. Their high melting points are also suggestive of ionic structures, and the molten salts also conduct electricity. Perhaps the most conclusive evidence is from X-ray diffraction studies, in which these compounds show uniform cation-anion distances. If they were molecular species, the interatomic distances within a molecule should be smaller than the interatomic distances between molecules.

7-2 LiBr has a formula weight of 86.845, and the unit cell contains four cations and four anions (or four formula units per molecular unit cell).

$$\frac{86.845 \text{ g mol}^{-1}}{3.464 \text{ g cm}^{-3}} = 25.07 \text{ cm}^3 \text{ mol}^{-1} \times \frac{10^{-6} \text{ m}^3}{\text{cm}^3} = 2.507 \times 10^{-5} \text{ m}^3 \text{ mol}^{-1}$$

$$\frac{2.507 \times 10^{-5} \text{ m}^3 \text{ mol}^{-1}}{6.022 \times 10^{23} \text{ units mol}^{-1}} \times \frac{4 \text{ units}}{\text{unit cell}} = \frac{1.665 \times 10^{-28} \text{ m}^3}{\text{unit cell}}$$

$$\sqrt[3]{1.665 \times 10^{-28} \text{ m}^3} = 5.502 \times 10^{-10} \text{ m} = \text{unit cell length}$$

$2\,(r_+ + r_-) = 5.502 \times 10^{-10}$ m ; $r_+ + r_- = 2.751 \times 10^{-10}$ m = 275.1 pm

The sum of the ionic radii from Appendix B-1 is 272 pm.

7-3 CsCl has 8 Cl^- at the corners of the unit cell cube, with Cs^+ at the center. $r_+/r_- = 188/167 = 1.13$.
CaF_2 has the same structure in a single cube of F^- ions, but only half the cubes contain Ca^{2+}. $r_+/r_- = 126/119 = 1.06$. Both should have coordination number = 12 based on the radius ratios.

7-4 In the table below, CaF_2 is considered to have a fluoride ion in the body center of the overall unit cell and calcium ions in the body centers of the subunits (labeled internal here).

Compound	Corners	Edges	Face centers	Body centers	Internal	Total	Type
NaCl cations	8 × 1/8		6 × 1/2			4	MX
NaCl anions		12 × 1/4		1 × 1		4	MX
CsCl cations	8 × 1/8					1	MX
CsCl anions				1 × 1		1	MX
CaF_2 cations					4 × 1	4	MX_2
CaF_2 anions	8 × 1/8	12 × 1/4	6 × 1/2	1 × 1		8	MX_2

7-5 The unit cell dimension is $2r$, the volume is $8\ r^3$. Since this cell contains one molecule whose volume is $4/3\ \pi\ r^3$, the fraction occupied is $\dfrac{4/3\pi r^3}{8r^3} = 0.524 = 52.4\%$

7-6 Figure 7-8 shows the zinc blende unit cell, which contains four S atoms (net) in a fcc lattice and four Zn atoms in the body centers of the alternate smaller cubes. The diagrams below are each a view of two layers of such a cell, with ● indicating a Zn atom in this layer and ○ indicating a S atom in the layer below. The next pair of layers either above or below these has the opposite pattern, and the third repeats the original. The fcc lattice pattern can be seen for the Zn atoms (four corners and face center on the top face, four face centers in the middle layer, and four corners and face center on the bottom face). Each S atom (and each Zn atom) has two nearest neighbors in the layer above and two in the layer below, in the arrangement for a tetrahedral hole. Extending the patterns below shows the S fcc lattice.

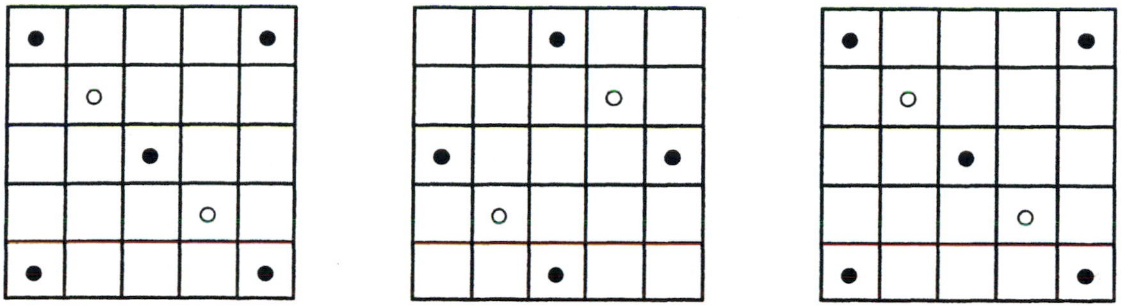

7-7 The graphite layers have essentially the same energy levels as benzene, but each level becomes a wider band because of the large number of atoms. This leads to the energy levels shown at the right, with the bands coming from the σ and π orbitals filled and those coming from the σ* and π* orbitals empty. The difference between the π and π* bands is small enough to allow conduction electrons to make the jump and the electrons and holes to move within the bands. Conduction perpendicular to the layers is smaller, because there are no direct orbitals connecting them. In polycrystalline graphite, the overall conductance is an average of the two.

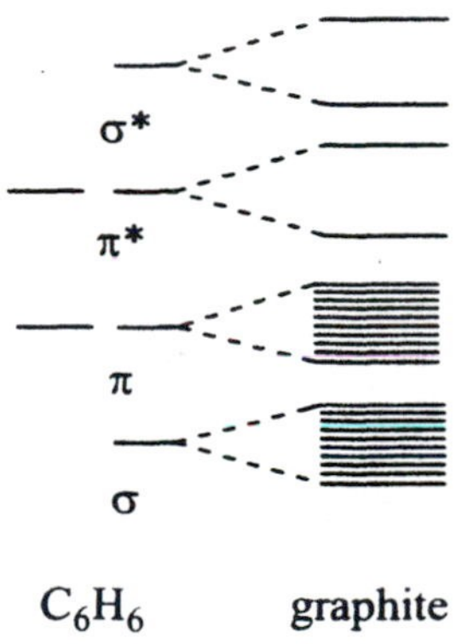

In diamond, each carbon atom has four σ bonds to its nearest neighbors. The gap between these filled orbitals and the corresponding antibonding orbitals is larger, effectively limiting conductance.

7-8 The unit cell length for a primitive cubic cell is $2\ r$. Using the Pythagorean Theorem, we can calculate the face diagonal as $\sqrt{(2r)^2 + (2r)^2} = 2.828r$ and the body diagonal as $\sqrt{(2.828r)^2 + (2r)^2} = 3.464r$. Each corner atom contributes r to this distance, so the diameter of the body center is $1.464\ r$ and the radius is $0.732\ r$, 73.2% of the corner atom size.

7-9 Hg(I) appears in compounds as Hg_2^{2+} units. The $6s^1\,4f^{14}\,5d^{10}$ structure of Hg^+ forms σ and σ* molecular orbitals from the *s* atomic orbitals, allowing the two *s* electrons to pair in the σ orbital for a diamagnetic unit.

7-10

	½ Cl_2 (*g*) → Cl (*g*)	239/2 = 119.5	
	Cl (*g*) + e^- → Cl^- (*g*)	−EA	
	Na(*s*) → Na(*g*)	109	
	Na (*g*) → Na^+ (*g*) + e^-	496	(= 5.14 eV × 96.4853 kcal/mol/eV)
	Na^+ (*g*) + Cl^- (*g*) → NaCl (*s*)	−772	
Total:	Na (*s*) + ½ Cl_2(*g*) → NaCl (*s*)	−413	

EA = 366 kJ/mol

$$U = \frac{NMZ_+Z_-}{r_0}\left[\frac{e^2}{4\pi\varepsilon_0}\right]\left(1 - \frac{\rho}{r_0}\right) = \frac{6.022\times10^{23}\times1.74756\times1\times(-1)}{281\times10^{-12}\,\text{m}}\times2.3071\times10^{-28}\,\text{J m}\times\left(1-\frac{30}{281}\right)$$

$$= -772 \text{ kJ/mol}$$

7-11 CaO has charges of 2+ and 2− and radii of 114 and 126 pm, total distance of 240 pm. KF has charges of 1+ and 1− and radii of 152 and 119 pm, total distance of 271 pm. The distance in CaO is 13% smaller and the charge factor is four times as large, both leading to larger interionic attraction and contributing to the hardness of the crystal.

MgO has charges of 2+ and 2− and radii of 86 and 126 pm, total distance of 212 pm, with NaCl structure and Madelung constant = 1.75.
CaF_2 has charges of 2+ and 1− and radii of 126 and 119 pm, total distance of 245 pm with the fluorite structure and Madelung constant = 2.52. The size difference and charges favor larger MgO interionic attraction, enough to overcome the Madelung constant difference.

7-12

Cation radius (pm) calculated from interionic distance and anion radius

	Li	Na	Ag
F (119 pm)	82	112	127
Cl (167 pm)	90	114	110
Br (182 pm)	93	116	106
r_+ from App. B-1	90	116	129

Hard-soft combinations (LiBr, AgF) have larger distances than calculated. Hard-hard (LiF) and soft-soft (AgBr) have smaller. NaF is a hard-hard combination by this criterion. The radius ratios for NaF and Ag F are large enough to predict CN = 8, the CsCl structure. All the others fit the CN = 6, NaCl structure criteria.

7-13 $MgCl_2$ has a rutile structure, so it will be used for the $NaCl_2$ calculation.

The lattice energy is for a rutile structure with combined radii of 270 pm (either $NaCl_2$ or $MgCl_2$) is

$$U = \frac{NMZ_+Z_-}{r_0}\left(\frac{e^2}{4\pi\varepsilon_0}\right)\left(1 - \frac{\rho}{r_0}\right)$$

$$U = \frac{6.022\times10^{23}\,\text{mol}^{-1}\times2.385\times2\times(-1)}{270\times10^{-12}\text{m}}\times2.307\times10^{-28}\,\text{J m}\left(1 - \frac{30}{270}\right)$$

$$= -2{,}182\ \text{kJ mol}^{-1}$$

The overall energy of formation of $MgCl_2$ is negative. For this calculation, the only differences between $MgCl_2$ and $NaCl_2$ are in the vaporization and ionization of the metal. The big difference is in the second ionization energy. $Na^+ \rightarrow Na^{2+} + e^-$, removing an electron from a closed $2p$ orbital, requires much larger energy than the corresponding $Mg^+ \rightarrow Mg^{2+} + e^-$ reaction that removes the second $3s$ electron.

In fact, the two metals differ by about 2,800 kJ/mol in these steps (all in kJ/mol):

Na (*s*) → Na (*g*)	107	Mg (*s*) → Mg (*g*)	147
Na (*g*) → Na^+ (*g*) + e^-	495	Mg (*g*) → Mg^+ (*g*) + e^-	738
Na^+ (*g*) → Na^{2+} (*g*) + e^-	4,562	Mg^+ (*g*) → Mg^{2+} (*g*) + e^-	1,451
Totals: Na (*s*) → Na^{2+} (*g*) + 2 e^-	5,164	Mg (*s*) → Mg^{2+} (*g*) + 2 e^-	2,336

Therefore, it is extremely unlikely that $NaCl_2$ can be made.

Repeating the process for a sodium chloride lattice with the sum of the radii 283 pm to compare the lattice energy and the energies necessary to form Na^+ and Mg^+:

$$U = \frac{NMZ_+Z_-}{r_0}\left(\frac{e^2}{4\pi\varepsilon_0}\right)\left(1 - \frac{\rho}{r_0}\right)$$

$$U = \frac{6.022\times10^{23}\,\text{mol}^{-1}\times1.748\times1\times(-1)}{283\times10^{-12}\text{m}}\times2.307\times10^{-28}\,\text{J m}\left(1 - \frac{30}{283}\right)$$

$$= -767\ \text{kJ mol}^{-1}$$

Na (*s*) → Na (*g*)	107	Mg (*s*) → Mg (*g*)	147
Na (*g*) → Na^+ (*g*) + e^-	495	Mg (*g*) → Mg^+ (*g*) + e^-	738
Totals: Na (*s*) → Na^+ (*g*) + e^-	602	Mg (*s*) → Mg^+ (*g*) + e^-	885

In this case, formation of MgCl seems possible, but formation of $MgCl_2$ is so much more favorable (nearly triple the lattice energy) that it is unlikely for it to stop at the MgCl stage.

7-14

$\frac{1}{2} Br_2(l) \rightarrow \frac{1}{2} Br_2(g)$	14.9
$\frac{1}{2} Br_2(g) \rightarrow Br(g)$	190.2
$Br(g) + e^- \rightarrow Br^-$	-324.7
$K(s) \rightarrow K(g)$	81.3
$K(g) \rightarrow K^+ + e^-$	418.8
$K^+ + Br^- \rightarrow KBr(s)$	-661.8
Total	-281.3 kJ mol^{-1} for $K(s) + \frac{1}{2} Br_2(l) \rightarrow KBr(s)$

For a sodium chloride lattice with total radii of 334 pm:

$$U = \frac{NMZ_+Z_-}{r_0}\left(\frac{e^2}{4\pi\varepsilon_0}\right)\left(1 - \frac{\rho}{r_0}\right)$$

$$U = \frac{6.022\times10^{23}\,\text{mol}^{-1}\times1.748\times1\times(-1)}{334\times10^{-12}\text{m}}\times2.307\times10^{-28}\text{J m}\left(1 - \frac{30}{334}\right)$$

$$= -661.8 \text{ kJ mol}^{-1}$$

7-15

$\frac{1}{2} O_2(g) \rightarrow O(g)$	247
$O(g) + 2e^- \rightarrow O^{2-}$	603
$Mg(s) \rightarrow Mg(g)$	37
$Mg(g) \rightarrow Mg^{2+} + 2e^-$	2188
$Mg^{2+} + O^{2-} \rightarrow MgO(s)$	-3932
Total	-857 kJ mol^{-1} for $Mg(s) + \frac{1}{2} O_2(g) \rightarrow MgO(s)$

For a sodium chloride lattice with total radii of 212 pm:

$$U = \frac{NMZ_+Z_-}{r_0}\left(\frac{e^2}{4\pi\varepsilon_0}\right)\left(1 - \frac{\rho}{r_0}\right)$$

$$U = \frac{6.022\times10^{23}\,\text{mol}^{-1}\times1.748\times2\times(-2)}{212\times10^{-12}\text{m}}\times2.307\times10^{-28}\text{J m}\left(1 - \frac{30}{212}\right)$$

$$= -3934 \text{ kJ mol}^{-1}$$

7-16 Radius ratios for the alkali halides:

Ions	Radii	Li^+	Na^+	K^+	Rb^+	Cs^+
Radii		90	116	152	166	206
F^-	119	0.76	0.98	1.28	1.40	1.73
Cl^-	167	0.54	0.70	0.91	0.99	1.23
Br^-	182	0.49	0.64	0.84	0.91	1.13
I^-	206	0.44	0.56	0.74	0.81	1.00

Those with a light outline have radius ratios indicating a coordination number of 8, and those with a dark outline indicate a coordination number of 12. All of these compounds actually have the sodium chloride lattice (CN = 6), so only 6 of the 20 give the correct predicted structure, with KI also very close to the CN = 6 ratio.

7-17 **a.** Forming anions from neutral atoms results in addition of an electron. More electrons means a larger size, due to electron-electron repulsion. Forming cations from neutral atoms results in removal of an electron and smaller size because of the same argument. The cations have fewer electrons and less electron-electron repulsions.

b. The oxide ion is larger than the fluoride ion because its nuclear charge is smaller, reducing the attraction for the electrons.

7-18 PbS has total radii of 303 pm. For the NaCl lattice:

$$U = \frac{NMZ_+Z_-}{r_0}\left(\frac{e^2}{4\pi\varepsilon_0}\right)\left(1 - \frac{\rho}{r_0}\right)$$

$$U = \frac{6.022\times 10^{23}\,\text{mol}^{-1}\times 1.748\times 2\times(-2)}{303\times 10^{-12}\,\text{m}}\times 2.307\times 10^{-28}\,\text{J m}\left(1 - \frac{30}{303}\right)$$

$$= -2{,}888\ \text{kJ mol}^{-1}$$

1/8 $S_8\,(s) \rightarrow S^{2-}\,(g)$	535
$Pb\,(s) \rightarrow Pb(g)$	196
$Pb\,(g) \rightarrow Pb^+ + e^-$	716
$Pb^+ \rightarrow Pb^{2+} + e^-$	1450
$Pb^{2+} + S^{2-} \rightarrow PbS\,(s)$	U

Total -23 kJ mol^{-1} for $Pb\,(s) + {}^1/_8\,S_8(s) \rightarrow PbS\,(s)$1

U = -2,386 kJ mol^{-1}, a difference of 23 %.

7-19 In ZnO or TiO, additional Zn or Ti would have two more electrons than the metallic ions. As a result, any nonstoichiometry in the direction of excess Zn or Ti would supply extra electrons, making an *n*-type semiconductor.

In Cu_2S, CuI, or ZnO, excess S, I, or O would have fewer electrons than the corresponding ions. Therefore, the result of excess nonmetals in the lattice would be a *p*-type semiconductor.

7-20 Vibrational motions of the atoms in the lattice become at least partly synchronized, with positive centers moving closer together. This concentration of positive charge can attract electrons, allowing two electrons to be closer to each other than would usually be the case. When the vibrations are synchronized, this attraction can ripple through the material, helping the electrons move. Apparently the whole system acts as if it is at ground state energies, so no net change in energy is needed to keep the process going indefinitely.

7-21   The general reaction is $Na_2Z + Ca^{2+}$ ($(aq) \rightarrow CaZ + 2\,Na^+$ (aq), where Na_2Z is the original zeolite with sodium ions providing the positive charge. When hard water, containing Ca^{2+} or Mg^{2+}, passes through the zeolite, the ions exchange, leaving only Na^+ cations in the softened water. The zeolite can be regenerated by flushing with concentrated brine. The large Na^+ concentration reverses the reaction above.

7-22   The ion C_2^{4-} should have the following molecular orbitals (see figure 5-7):

$$
\begin{array}{llcc}
\sigma_u{}^*(2p) & \underline{\quad} & & \underline{\quad} \\
\pi_g{}^*(2p) & \underline{\uparrow}\ \ \underline{\uparrow} & & \underline{\quad}\ \ \underline{\uparrow\downarrow} \\
\\
\sigma_g(2p) & \underline{\uparrow\downarrow} & & \underline{\uparrow\downarrow} \\
\pi_u(2p) & \underline{\uparrow\downarrow}\ \ \underline{\uparrow\downarrow} & \longrightarrow & \underline{\uparrow\downarrow}\ \ \underline{\uparrow\downarrow} \\
\\
\sigma_u{}^*(2s) & \underline{\uparrow\downarrow} & & \underline{\uparrow\downarrow} \\
\sigma_g(2s) & \underline{\uparrow\downarrow} & & \underline{\uparrow\downarrow}
\end{array}
$$

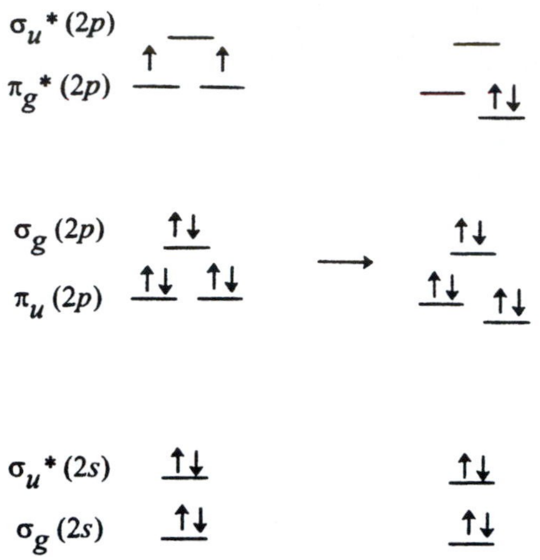

Distortion could result in removal of the degeneracy of the π_u and $\pi_g{}^*$ orbitals, giving a diamagnetic ion.

Chapter 8
Chemistry of the Main Group Elements

8-1 **a.**

H_2	74.2 pm	436 kJ/mol
H_2^+	106 pm	255 kJ/mol

Yes, these values are consistent with the molecular orbital descriptions. H_2 has two electrons in the bonding σ orbital, H_2^+ has only one. Therefore, the attraction between the bonding electron(s) and the nuclei are weaker in the case of H_2^+, and the bond distance is longer.

b. H_3^+ was described in problems 5-13 and 5-32. It has a single pair of electrons in a 2-electron-3-center bond, with bond orders of 1/3.

8-2 He_2^+ has two electrons in the bonding σ orbital and one in the antibonding σ* orbital, with a bond order of one-half. HeH^+ has a bond order of one, but with a poorer match between the energy levels on the two atoms (-13.6 eV for H, -24.5 for He).

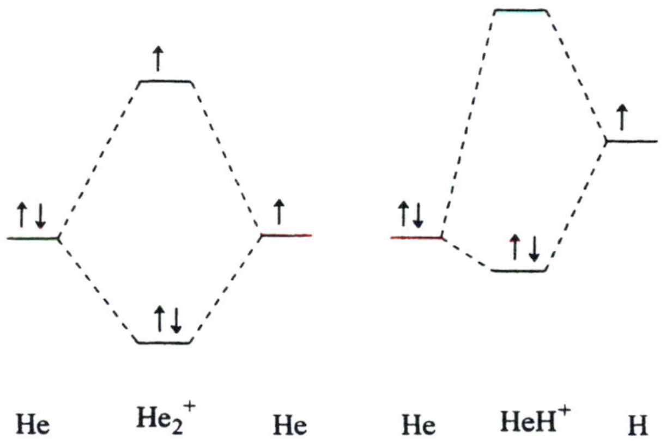

8-3 Figure 8-7 on page 252 shows the equilibrium constants for the alkali metal ions, with the sodium constant larger than either the lithium or potassium constants. A similar curve is found for the alkaline earths, with the maximum at strontium. Apparently the optimum size for an ion fitting in the cryptand is larger than sodium, but smaller than barium or potassium, leaving strontium as the closest fit for these two series.

Ionic radii, in pm:

Li^+	90	Ca^{2+}	114
Na^+	116	Sr^{2+}	132
K^+	149	Ba^{2+}	152

8-4 The diagram at the right shows the primary interactions forming molecular orbitals. The other orbitals on the fluorine atoms form lone-pair orbitals and π orbitals. These diagrams are oversimplified, as shown by the pictures below showing the same four molecular orbitals.

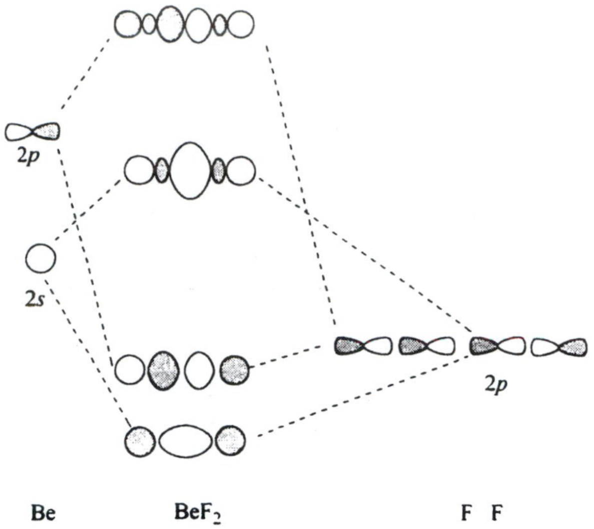

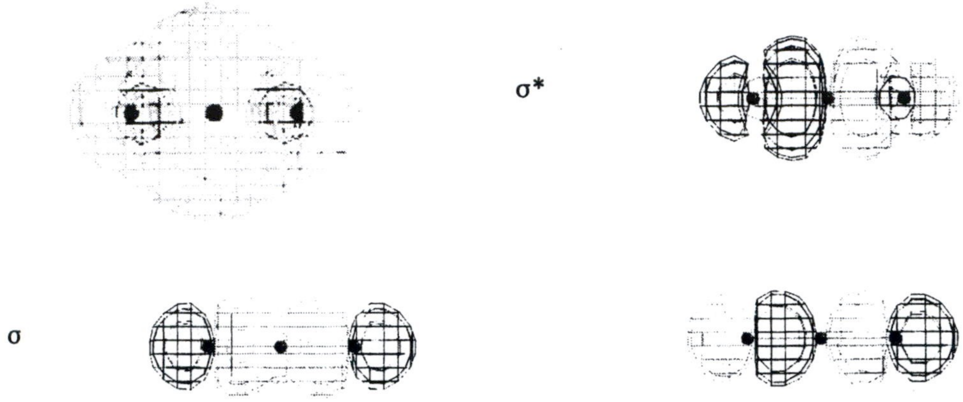

Orbitals using Be *s* orbital

Orbitals using Be *p* orbital

8-5 A combination of the in-plane *p* orbitals of Cl and the *s* and in-plane *p* orbitals of Be form the orbitals that link the Be and Cl atoms in the three-atom, two-electron bonds. The combinations that participate are shown on the next page, more than might be expected by the simple description. Bonding interactions are shown; the corresponding antibonding orbitals are also formed, in addition to others primarily involved with the terminal chlorine atoms.

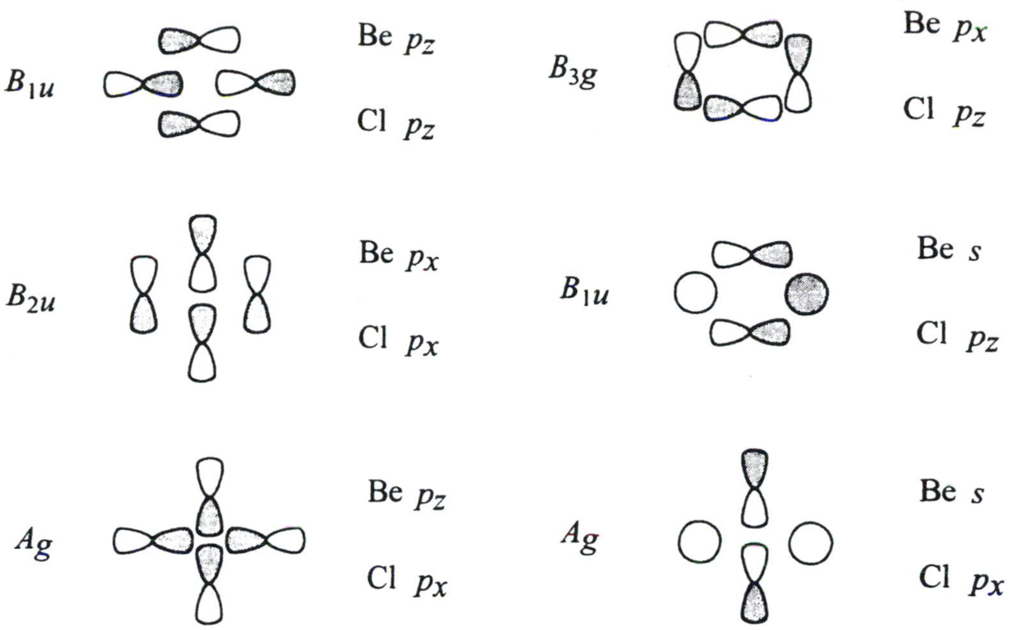

The same orbitals, shown as calculated by the AM1 option of Spartan Plus, are shown below. Orbitals from the terminal chlorine atoms are included in these molecular orbitals.

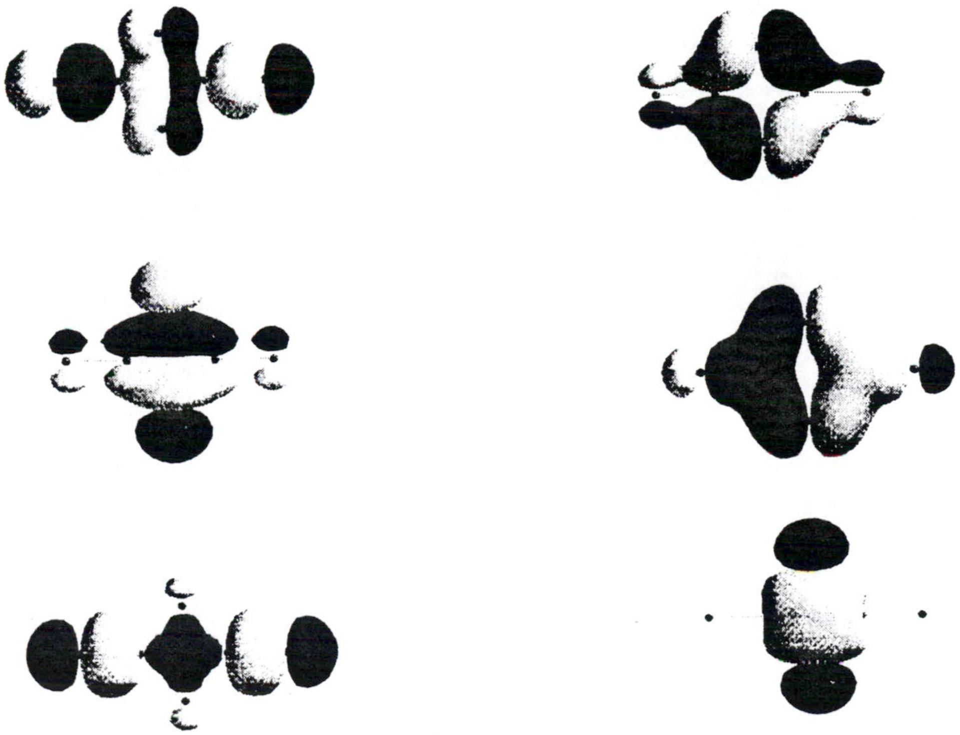

8-6 The much greater difference in orbital energies of B (-8.3 and -14.0 eV) and F (-18.7 and -40.2 eV) makes the BF bond weaker than that of CO. The *s* orbital of B and the *p* orbital of F can interact, but the *p-p* combination cannot. This mismatch results in weak single bonding, rather than the triple bond of CO.

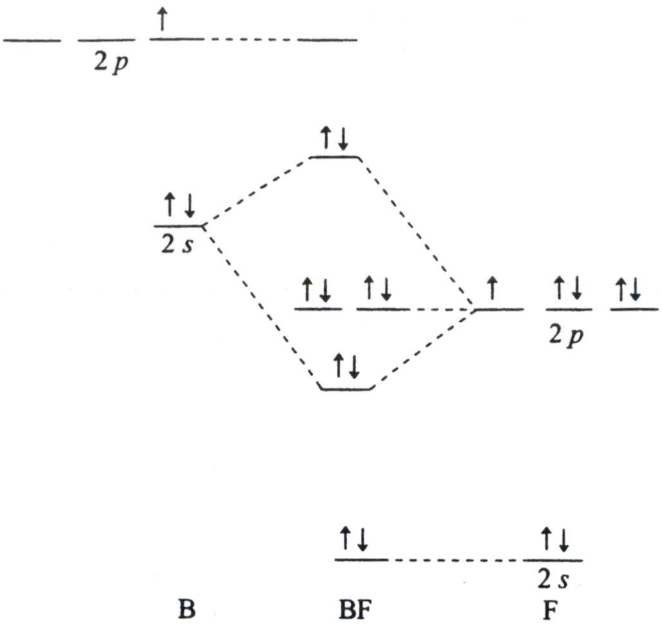

8-7 The combination of orbitals forming the bridging orbitals in $Al_2(CH_3)_6$ is similar to those of diborane (pp. 256-258). The difference is that the CH_3 group has a *p* orbital (or sp^3 hybrid orbital) available for bonding to the aluminum atoms, with the same symmetry possibilities as those of the H atoms in B_2H_6.

8-8

D_{2h}	E	$C_2(z)$	$C_2(y)$	$C_2(x)$	i	$\sigma(xy)$	$\sigma(xz)$	$\sigma(yz)$	
$\Gamma(p_z)$	2	2	0	0	0	0	2	2	
$\Gamma(p_x)$	2	-2	0	0	0	0	2	-2	
$\Gamma(1s)$	2	0	0	2	0	2	2	0	
A_g	1	1	1	1	1	1	1	1	z^2
B_{2g}	1	-1	1	-1	1	-1	1	-1	xz
B_{1u}	1	1	-1	-1	-1	-1	1	1	xy
B_{3u}	1	-1	-1	1	-1	1	1	-1	x

All the following can be seen by direct addition or by following the procedure for reducing a reducible representation. The first two are for boron orbitals, the third for hydrogen orbitals.

a. $\Gamma(p_z) = A_g + B_{1u}$

b. $\Gamma(p_x) = B_{2g} + B_{3u}$

c. $\Gamma(1s) = A_g + B_{3u}$

d. Treating each group orbital as a single orbital, the orbitals below have the indicated symmetries.

D_{2h}		E	$C_2(z)$	$C_2(y)$	$C_2(x)$	i	$\sigma(xy)$	$\sigma(xz)$	$\sigma(yz)$
$A_g(p_z)$		1	1	1	1	1	1	1	1
$B_{1u}(p_z)$		1	1	-1	-1	-1	-1	1	1
$B_{2g}(p_x)$		1	-1	1	-1	1	-1	1	-1
$B_{3u}(p_x)$		1	-1	-1	1	-1	1	1	-1
$A_g(s)$		1	1	1	1	1	1	1	1
$B_{3u}(s)$		1	-1	-1	1	-1	1	1	-1

8-9 Carbon has two lone pairs in $C(PPh_3)_2$, with VSEPR predicting a tetrahedral angle. The bulky phenyl groups force a larger angle.

8-10 Comparing the compounds CaC_2, CeC_2, and YC_2, the unusual 2+ group 3 ions each have an extra electron. It has been suggested (Greenwood and Earnshaw, *Chemistry of the Elements,* 2nd ed., pp. 299) that there is some transfer of this extra electron to the π^* orbitals of the dicarbide (or acetylide) ion, resulting in a longer bond.

8-11 Radioactive decay obeys a first order kinetic equation:

$$\frac{dx}{dt} = -kx; \quad \ln\left(\frac{x}{x_0}\right) = -kt$$

The relationship of the half-life (the time at which $x = 1/2\, x_0$) and the rate constant is

$$\ln\left(\frac{x}{x_0}\right) = \ln\left(\frac{1}{2}\right) = -kt_{1/2} = -0.693$$

$$k = \frac{0.693}{t_{1/2}}$$

$$k = \frac{0.693}{t_{1/2}} = 1.21 \times 10^{-4}\, y^{-1}$$

$$\ln\left(\frac{x}{x_0}\right) = \ln(.56) = -1.21 \times 10^{-4}\, t$$

$$t = 4.8 \times 10^3\, y$$

8-12 I_h symmetry includes 12 C_5 axes, 12 C_5^2 axes, 20 C_3 axes, 15 C_2 axes, an inversion center, 12 S_{10} axes, 20 S_6 axes, and 15 mirror planes. One of the C_5 axes and an S_{10} axis can be seen in the middle of the second fullerene figure (end view) in Figure 8-16. There are six of each of these, each representing two rotation axes, C_5 and C_5^4, for a total of 12. The same axes fit the 12 S_5^2 axes. If the ball is rotated to line up two hexagons surrounded by alternating pentagons and hexagons, the 20 C_3 and 20 S_6 axes can be seen (one in the center, three in the first group around that hexagon, and three adjacent pairs in the next group, doubled because of C_3 and C_3^2). The C_2 axes pass through bonds shared by two hexagons, with pentagons at each end. These original hexagons each have two more, making 5, the next ring out has 8, and the perimeter (seen edge on) has two for a total of 20 C_2. There are five mirror planes through the center of the pentagons surrounded by five hexagons (see the end view of Figure 8-16 again), and there are three sets of these for a total of 15.

8-13 **a.** The symmetry of C_{70} is D_{5h}.

b. If two of these are joined as described, a mirror plane perpendicular to the two new bonds is the only symmetry element. C_s. Other point groups are possible for linked dimers - see the reference for these.

8-14 The increased stability of 2+ oxidation states as compared to 4+ is an example of the "inert pair" effect (see Greenwood and Earnshaw, *Chemistry of the Elements*, 2nd ed., pp. 226, 227, 374). In general, the ionization energy decreases going down a column of the periodic table, because of greater shielding by the inner electrons. In this family, removal of the second electron is fairly easy, as it is the first in the higher energy *p* orbitals. However, the next two electrons to be removed are the *s* electrons, and they are not as thoroughly shielded in the ions. The effect is larger for the three lower members of the group because the d^{10} electrons are added between the *s* and *p* electrons. The lower electronegativity of C and Si make them more likely to form covalent bonds than ions.

8-15 **a.** $H_2ISiSiH_2I$ has a C_2 axis perpendicular to the plane of the paper as drawn, a perpendicular mirror plane (the plane of the paper), and an inversion center. Overall, C_{2h}.

b. The Si-H stretches have the reducible representation shown below, which reduces to $\Gamma = A_g + B_g + A_u + B_u$. A_u and B_u are IR-active.

C_{2h}	E	C_2	i	σ_h	
Γ	4	0	0	0	
A_g	1	1	1	1	R_z
B_g	1	-1	1	-1	R_x, R_y
A_u	1	1	-1	-1	z
B_u	1	-1	-1	1	x, y

8-16 $P_4\ (g) \rightleftharpoons 2\ P_2\ (g)$, $\Delta H = 217$ kJ mol^{-1}
P_4 has six P–P bonds, so six bonds are broken and two triple bonds are formed.

$\Delta H = \Sigma$bond dissociation energy (react) – Σbond dissociation energy (prod)

$217 = 6 \times 200 - 2$(bond dissociation energy of P≡P)

bond dissociation energy of P≡P $= 492$ kJ mol^{-1}

The single P–P bonds are stronger than N–N bonds, probably because the *p* orbitals extend farther and overlap better to form the σ bond. The p_π orbitals in P_2 do not overlap as effectively as those in N_2, resulting in weaker π bonds in P_2.

8-17 **a.** N_3^- has molecular orbitals similar to those of CO_2 (pp. 143-147), with two occupied σ orbitals and two occupied π orbitals for a total of 4 bonds. Here the atomic orbitals have identical energies.

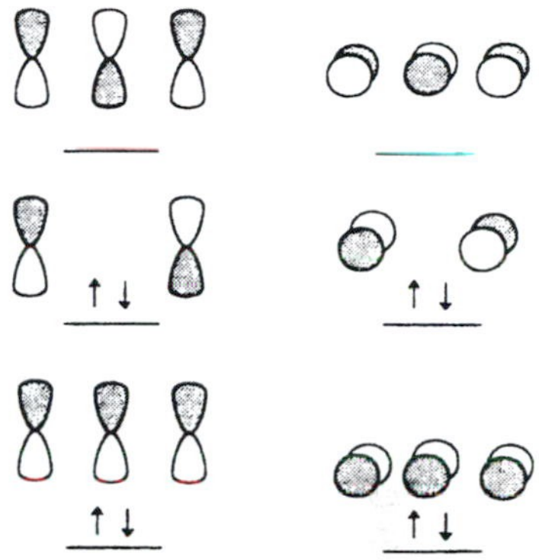

b. Because the HOMOs of N_3^- are primarily composed of *p* orbitals of the terminal nitrogen atoms, H^+ bonds at an angle to the N=N=N axis. The angle is 114°, larger than the simple 90° predicted by bonding to a *p* orbital.

c. The H–N–N angle is nearly that of sp^2 hybrids. Such bonding reduces the importance of the resonance structure with a triple bond at the H end of the molecule and increases the importance of the structure with the triple bond at the opposite end. In MO terms, the proton draws electron density away from the nearby N–N bond, weakening it and strengthening the bond at the opposite end of the molecule.

8-18 Hydrazine has two positive hydrogens on each nitrogen, ammonia has three. This leaves the ammonia nitrogen more negative, and therefore more basic.

8-19 The larger central atoms force the bonded pairs farther away from the central nucleus, reducing their repulsive force. The lone pair remains closer to the nucleus, and therefore has more influence and forces a smaller angle for the molecules with larger central atoms.

8-20 One *p* orbital of each oxygen is used for the σ bond to nitrogen and the *p* orbital perpendicular to the plane of the molecule is used for a π orbital. The remaining *p* orbital is in the plane of the molecule, so addition of a proton to this lone pair leaves the entire molecule planar.

8-21 From Table 8-10:

N_2O	$C_{\infty v}$	NO	$C_{\infty v}$	NO_2	C_{2v}	N_2O_3	C_s
N_2O_4	D_{2h}	N_2O_5	D_{2h}	NO^+	$C_{\infty v}$	NO_2^+	$D_{\infty h}$
NO_2^-	C_{2v}	NO_3^-	D_{3h}	$N_2O_2^{2-}$	C_{2h}	NO_4^{3-}	T_d
HNO_2	C_s	HNO_3	C_s				

8-22 S_2 is similar to O_2, with a double bond. As a result, the bond is shorter than single bonds in S_8.

8-23 MnF_6^{2-} acts as a Lewis base, donating two F^- to SbF_5 molecules:

$$MnF_6^{2-} + 2\ SbF_5 \rightarrow 2\ SbF_6^- + MnF_4$$

The remaining MnF_4 can then lose an additional F atom, with two combining to form F_2:

$$2\ MnF_4 \rightarrow F_2 + 2\ MnF_3$$

8-24 I_3^- is linear because there are three lone pairs in a trigonal geometry on the central I. I_3^+ is bent because there are only two lone pairs on the central I, leaving a structure similar to H_2O.

8-25 B has only three electrons. Adding six from the hydrogen atoms gives B_2H_6 a total of 12 electrons. The 3-center 2-electron bonds results in four pairs around each boron atom and nearly tetrahedral symmetry at each boron atom.
I has seven electrons initially. In I_2Cl_6, five more are added to each iodine, resulting in 12 electrons and octahedral symmetry around each iodine.

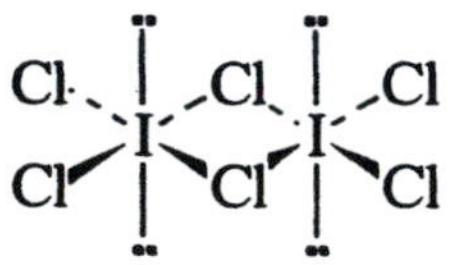

8-26 $F^- + BrF_3 \rightarrow BrF_4^-$ KF acts as a base, BrF_4^- is the solvent anion.
$SbF_5 + BrF_3 \rightarrow BrF_2^+ + SbF_6^-$ BrF_2^+ is the cation of the solvent; SbF_5 acts as the acid.

8-27 **a.** Br_2^+ and I_2^+ have each lost one antibonding electron, so they have bond orders of 1.5 and shorter bonds than the neutral molecules.

b. The HOMOs of I_2^+ and Br_2^+ are π^*, so the transitions are probably $\pi^* \rightarrow \sigma^*$, just as those of the neutral molecules. The energies are slightly different from those of the molecules; I_2^+ has closer spacing of its orbital levels, absorbing orange light compared to the green absorbed by Br_2^+.

8-28 There are 12 *p* orbitals on the four I atoms, forming 12 MOs, as shown here. Those in the center column form π orbitals, those in the outer columns form σ orbitals. Nine of these orbitals are filled by the 18 *p* electrons.

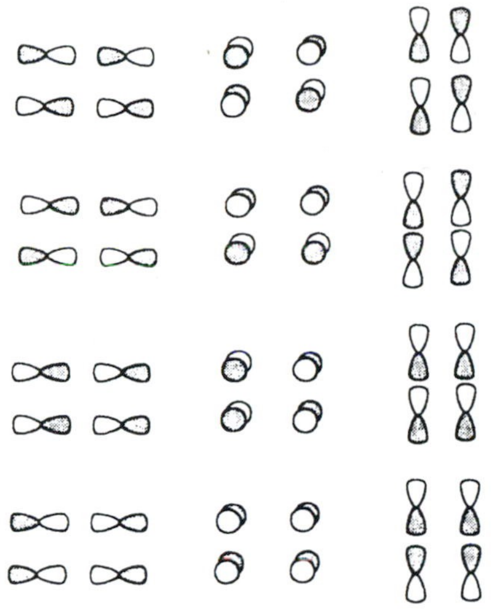

8-29 These reactions take place in the gas phase. The initial product of Xe with PtF_6 is believed to be Xe^+ PtF_6^-; however, when these two ions are in close proximity, they may react further to give $[XeF]^+$, $[Pt_2F_{11}]^-$, and other products. SF_6, if present in large excess, prevents the formation of these secondary products, apparently by acting as an inert diluent and preventing effective collisions between the desired products.

8-30 All these compounds have singly bonded F, doubly bonded O, and lone pairs, with 12 to 16 electrons around Xe, depending on the degree of double bonding used.

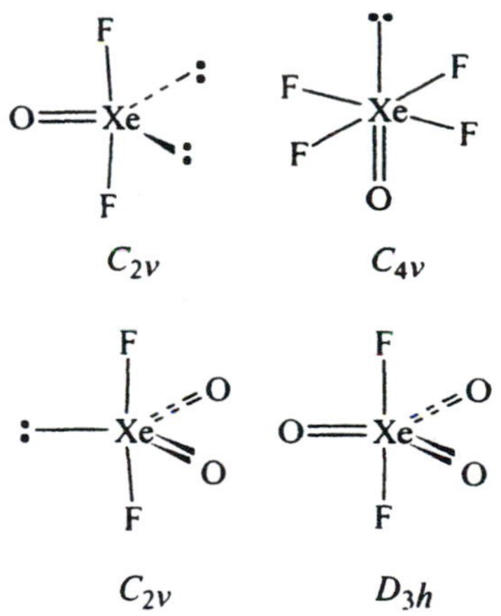

8-31 Two electrons are in the bonding orbital and two are in the nonbonding orbital:

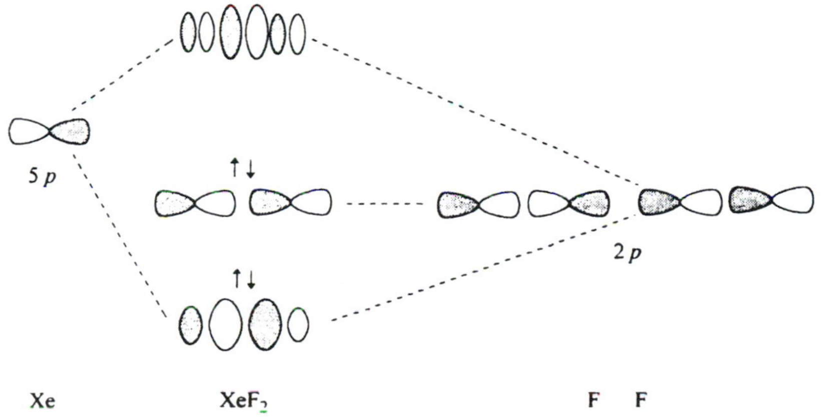

8-32 $Xe(OTeF_5)_4$: The $OTeF_5$ group has one electron available for bonding on the O, so the four groups form a square planar structure around Xe, with lone pairs in the axial positions. $O{=}Xe(OTeF_5)_4$: A square pyramidal structure, with O in one of the axial positions and a lone pair on the other, and $OTeF_5$ groups in the square base.

8-33 Half reactions:

$$Mn^{2+} + 4H_2O \rightarrow MnO_4^- + 8H^+ + 5e^-$$
$$XeO_6^{4-} + 12H^+ + 8e^- \rightarrow Xe + 6H_2O$$

Overall reaction:

$$8Mn^{2+} + 5XeO_6^{4-} + 2H_2O \rightarrow 8MnO_4^- + 4H^+ + 5Xe$$

8-34 XeF_5^- has D_{5h} symmetry. The reducible representation for Xe-F stretch is

	E	$2C_5$	$2C_5^2$	$5C_2$	σ_h	$2S_5$	$2S_5^3$	$5\sigma_v$
$\Gamma =$	5	0	0	1	5	0	0	1

which reduces to $\Gamma = A_1' + E_1' + E_2'$, with only E_1' IR-active.

8-35 $[FBeNe]^+$ The F-Be bonding has several orbitals contributing:

HOMO and HOMO-1 degenerate pair,
π bonding between Be and F

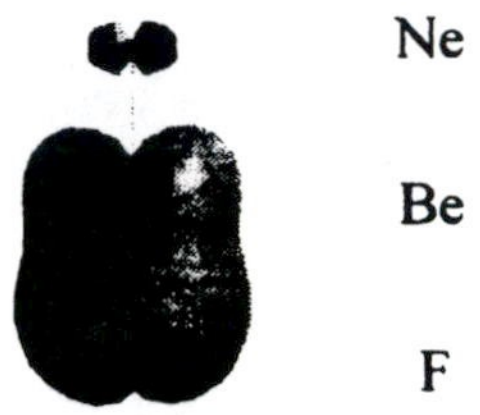

HOMO−2, σ bonding by the F p_z and the Be s

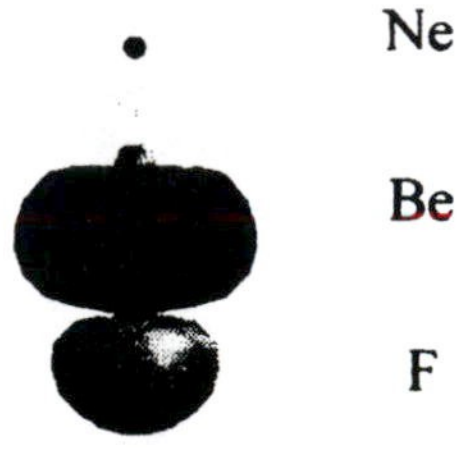

The Be-Ne bonding has only one molecular orbital contributing significantly:

HOMO−5, σ bonding by the Ne p_z and the Be s and p_z

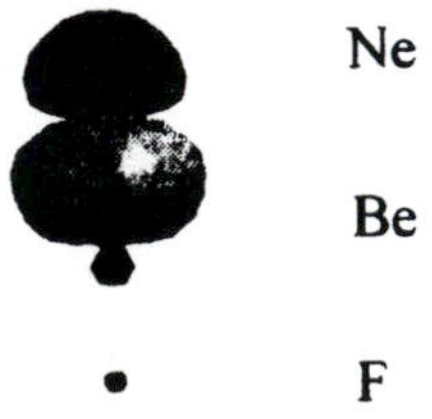

Overall, the bonding looks like $F{\equiv}Be{-}Ne$.

8-36 In Xe_2^+, there are a total of 54 orbitals. We will classify only the higher-energy orbitals here, starting with the highest energy. The orbitals used are the 5*s*, 5*p*, and 4*d*.

Orbital	Symmetry	AOs		Orbital	Symmetry	AOs
HOMO	σ*	p_z		HOMO-8 HOMO-9	δ*	$d_{x^2-y^2}$, d_{xy}
HOMO-1 HOMO-2	π*	p_x p_y		HOMO-10 HOMO-11	δ	$d_{x^2-y^2}$, d_{xy}
HOMO-3	σ	p_z		HOMO-12 HOMO-13	π*	d_{xz}, d_{yz}
HOMO-4 HOMO-5	π	p_x p_y		HOMO-14 HOMO-15	π	d_{xz}, d_{yz}
HOMO-6	σ*	*s*		HOMO-16	σ*	d_{z^2}
HOMO-7	σ	*s*		HOMO-17	σ	d_{z^2}

If this were the neutral Xe_2, there would be no bond, because every occupied bonding orbital would be offset by an occupied antibonding orbital. In this case, one electron is missing, so there is 1/2 bond, making for a long bond. In addition, Xe is a large atom, so the bond would naturally be long. Together, these factors make for the longest main group bond measured.

Overlap for the δ and δ* orbitals is very small in this case because there are filled 5*s* and 5*p* orbitals outside the 4*d* orbitals.

Chapter 9
Coordination Chemistry I: Structures and Isomers

9-1 **a.** dicyanotetra(methylisocyano)iron(0) or dicyanotetra(methylisocyano)iron(0) - both methods of naming result in the same name.

b. rubidium tetrafluoroargentate(III) or rubidium tetrafluoroargentate(1-)

c. *cis*- and *trans*-carbonylchlorobis(triphenylphosphine)iridium(I) or *cis*- and *trans*-carbonylchlorobis(triphenylphosphine)iridium(0)

9-2 **a.** Bis(en)Co(III)-μ-amido-μ-hydroxobis(en)Co(III) ion

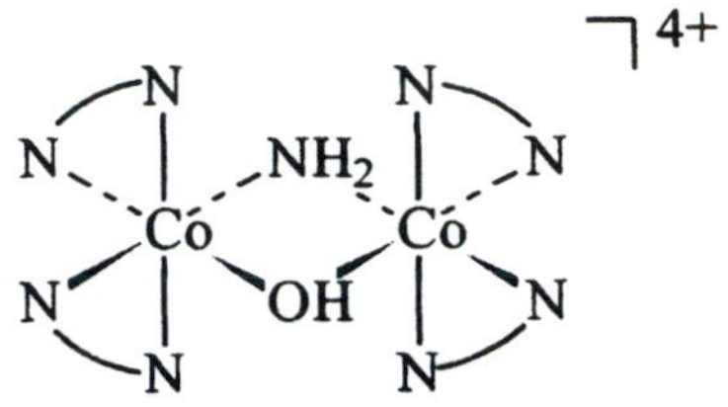

b. diaquadiiododinitrito Pd(IV)

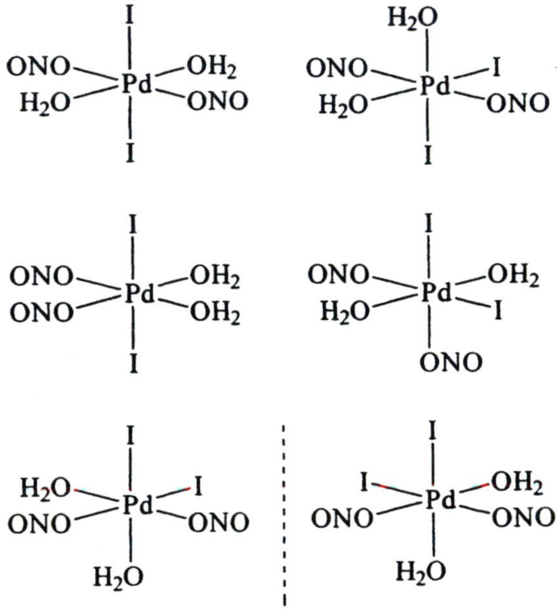

enantiomers

9-3 $M(AB)_3$

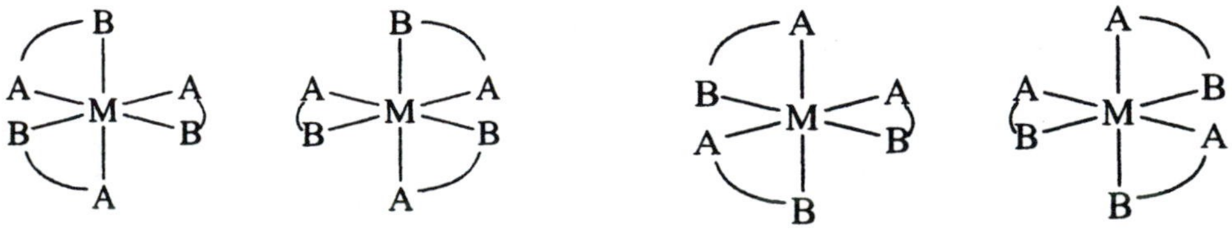

fac *mer*

9-4 **a.** pentaammineazidocobalt(III) sulfate or pentaammineazidocobalt(2+) sulfate

b. sodium tetrachloroaluminate(III) or sodium tetrachloroaluminate(1-)

c. carbonatobis(ethylenediamine)cobalt(III) chloride or carbonatobis(ethylenediamine)cobalt(1+) chloride

9-5 **a.** triammineaquadichlorocobalt(III) chloride

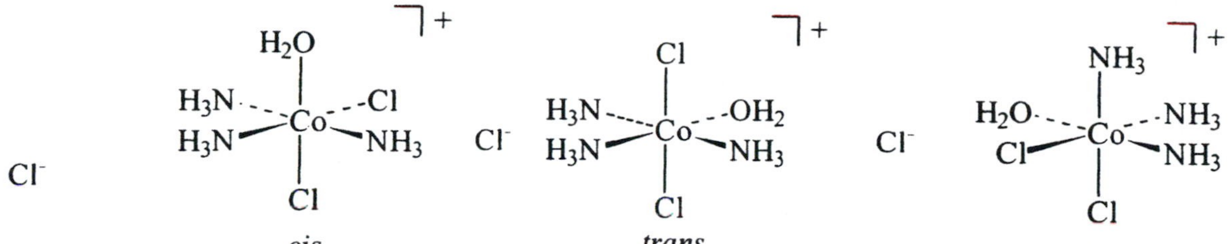

cis *trans* *fac*

mer

b. μ-oxo-bis(pentammine-chromium(III)) ion

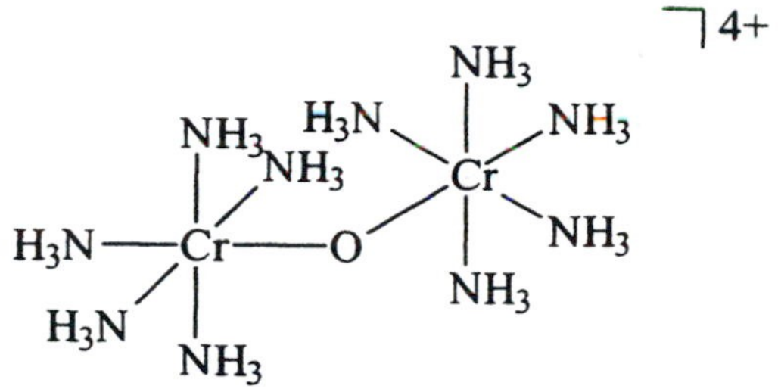

c. potassium diaquabis(oxalato)manganate(III)

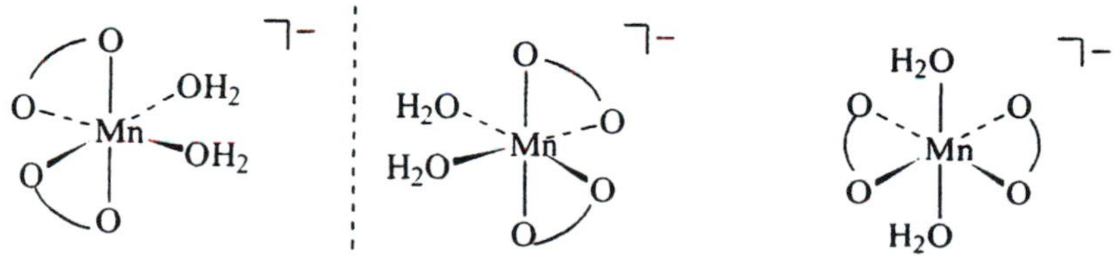

Λ *cis* Δ *trans*

9-6.

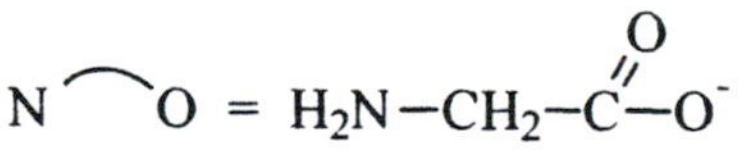

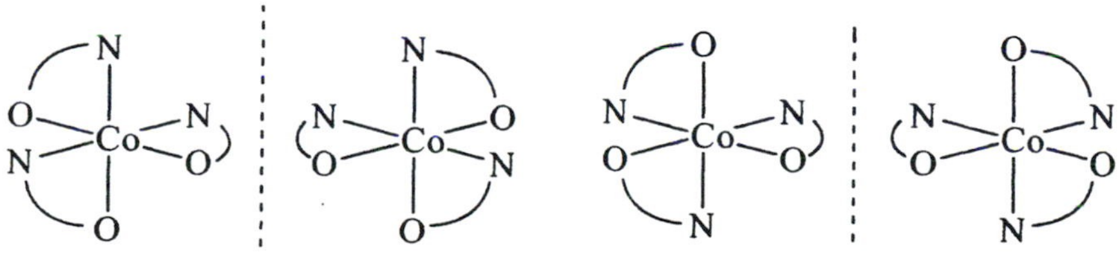

Λ *mer* Δ Δ *fac* Λ

9-7 I and II have only a C_2 axis through the middle of the N-Cl-N-Cl and P-Cl-P-Cl faces. C_2
III has only an inversion center. C_i

9-8 **a.** $[Pt(NH_3)_3Cl_3]^+$

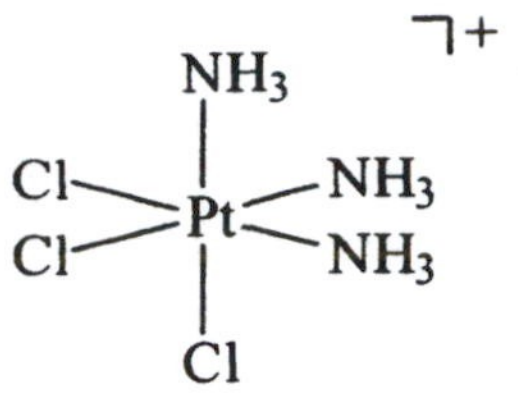

fac

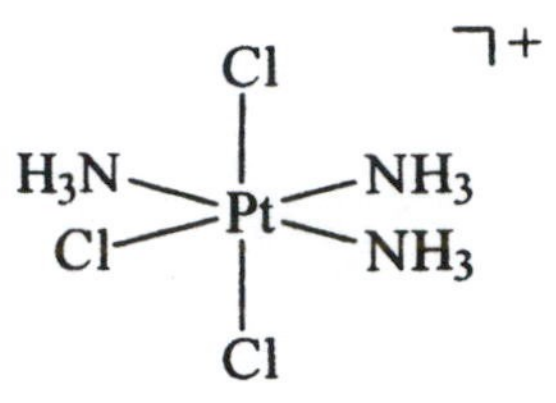

mer

b. $[Co(NH_3)_2(H_2O)_2Cl_2]^+$

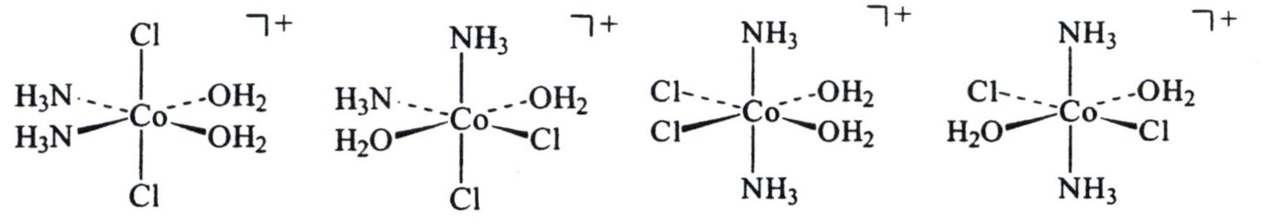

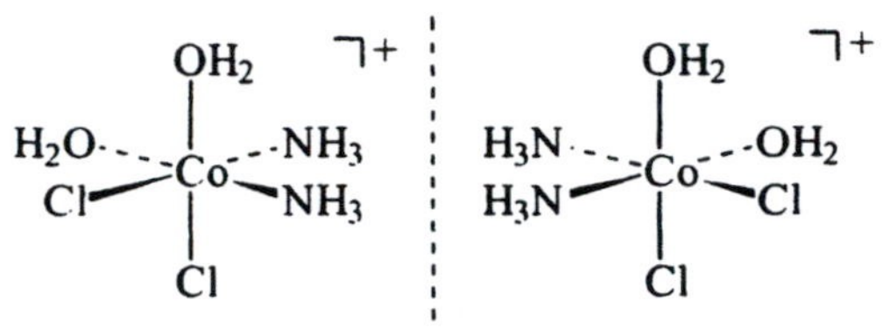

enantiomers

c. $[Co(NH_3)_2(H_2O)_2BrCl]^+$

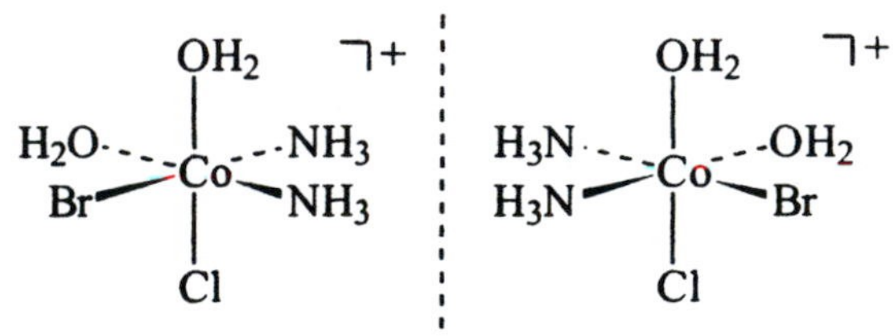

enantiomers

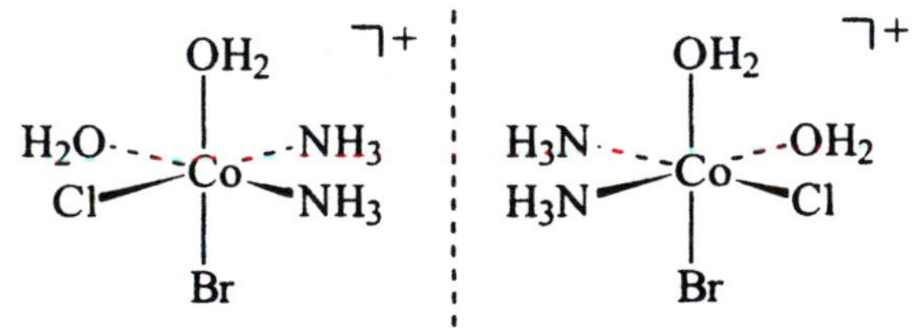

enantiomers

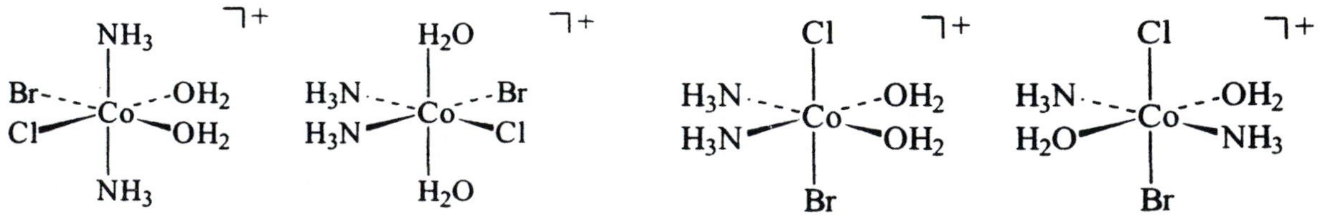

d. [$Cr(H_2O)_3BrClI$]

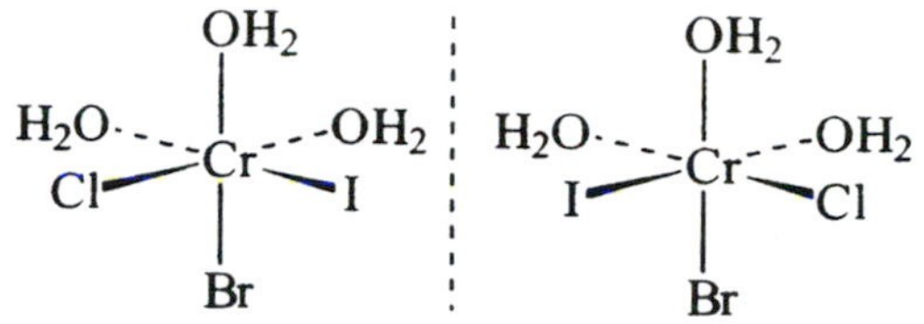

enantiomers

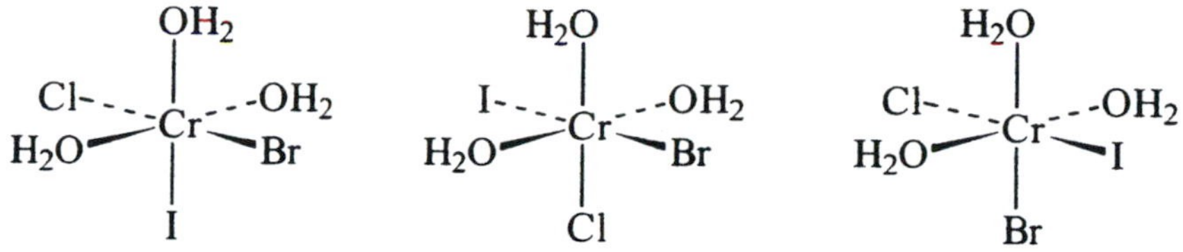

e. [$Pt(en)_2Cl_2$]$^{2+}$

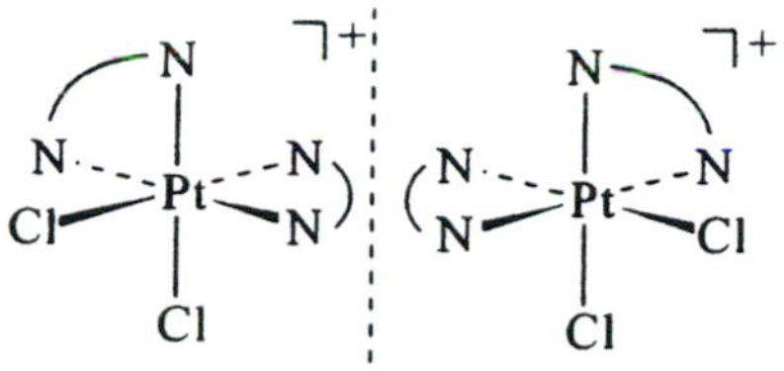

cis enantiomers

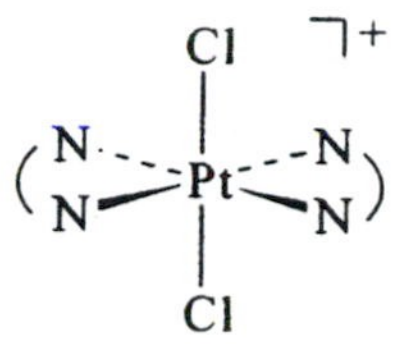

trans

f. [$Cr(o$-phen$)(NH_3)_2Cl_2$]$^+$

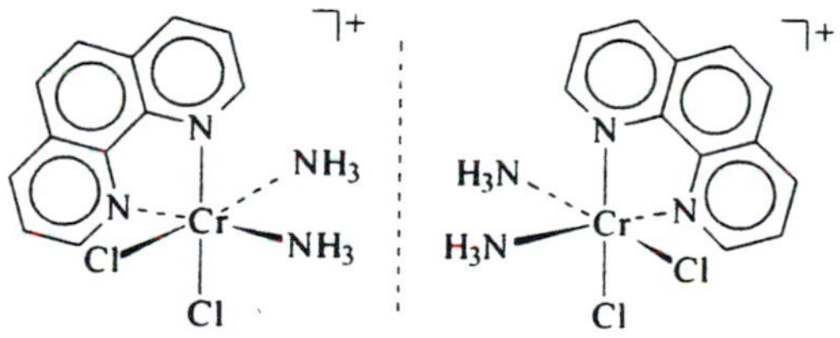

cis enantiomers

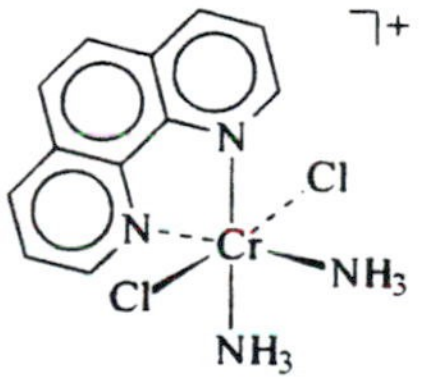

trans Cl's

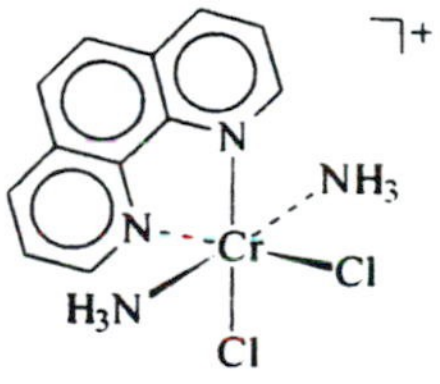

trans NH_3's

g. [$Pt(bipy)_2BrCl$]$^{2+}$

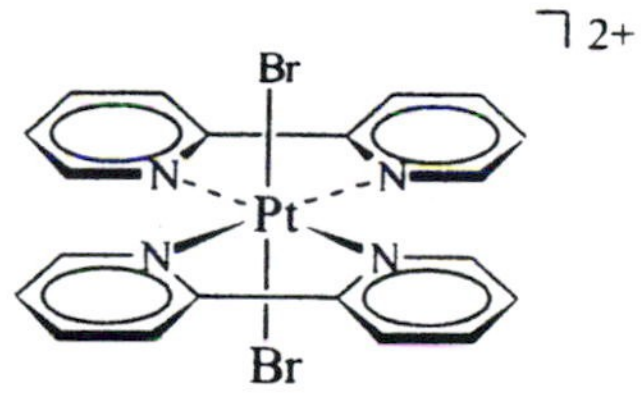

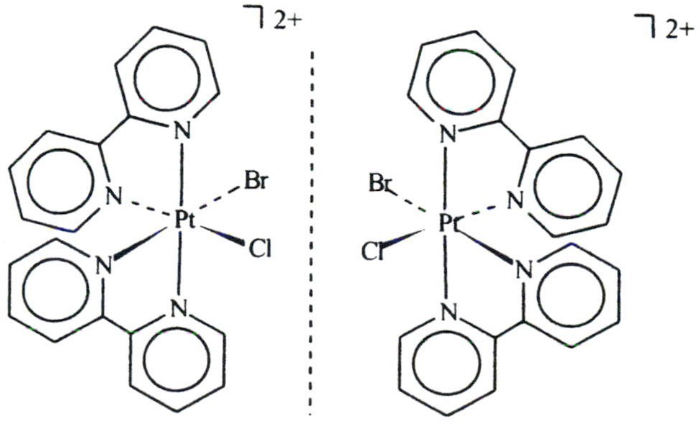

Λ Δ

h. $[Fe(dtc)_3]$

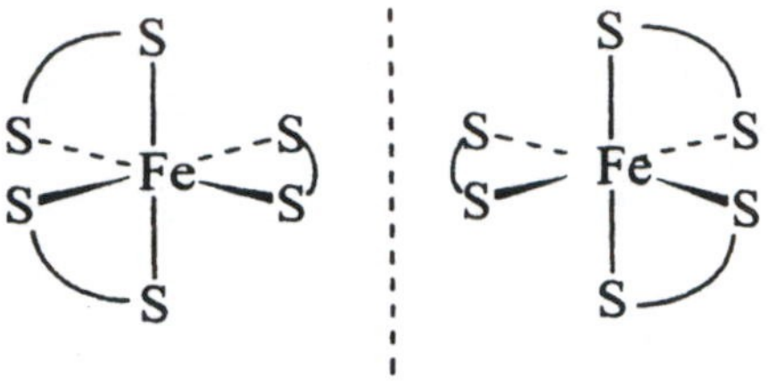

Λ Δ

i. $Re(arphos)_2Br_2]$

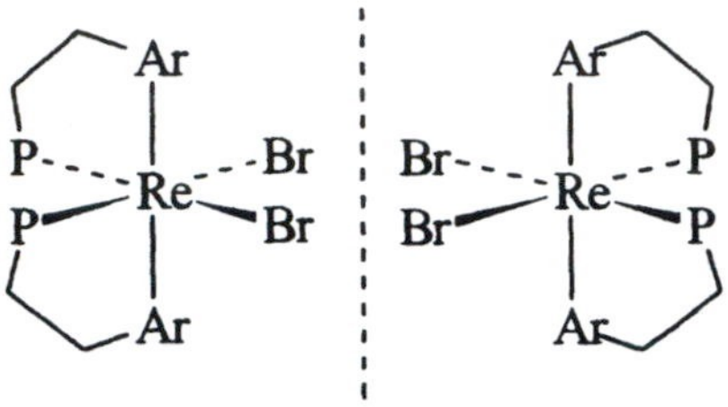

Λ Δ

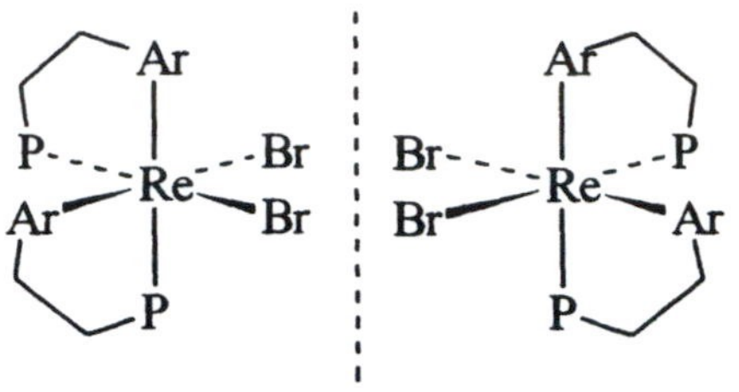

Λ Δ

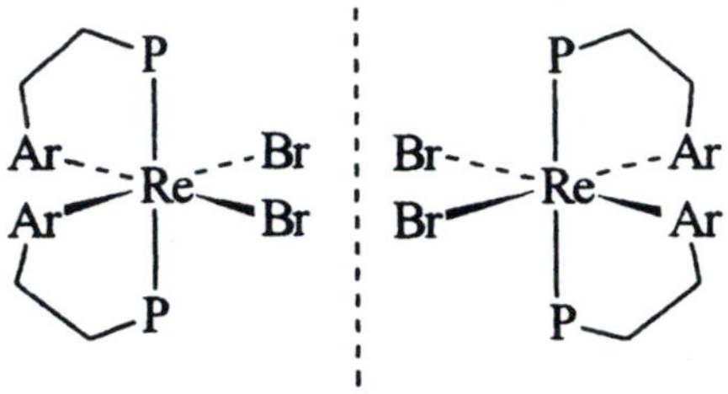

Λ Δ

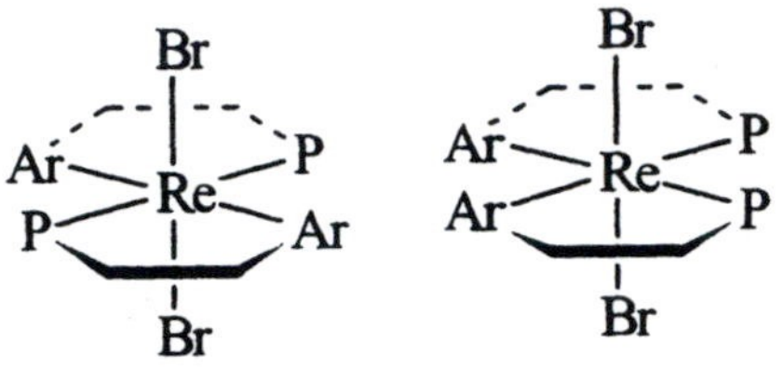

j. $[Re(dien)Br_2Cl]$

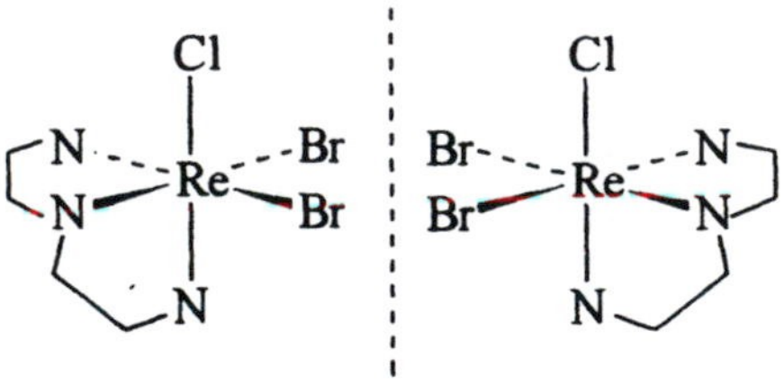

Λ Δ

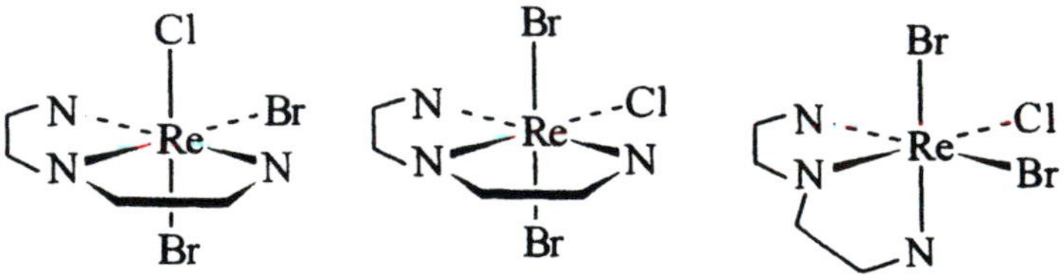

9-9 The single C–N stretching frequency indicates a *trans* structure for the cyanides, while the two C–O bands indicate a *cis* structure for the carbonyls. As a result, the bromides are also *cis*.

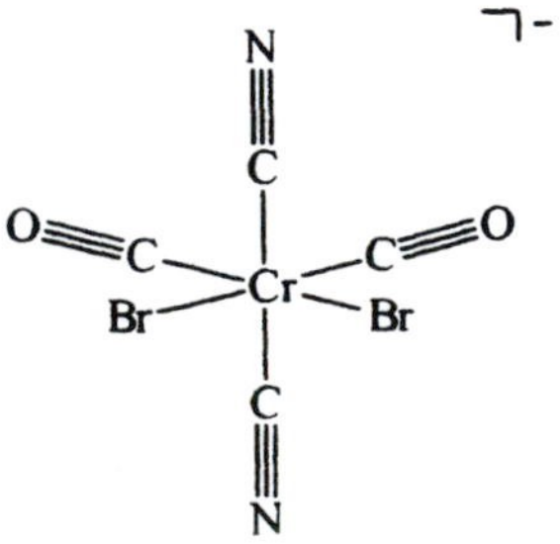

9-10 **a.** triamminotrichloroplatinum(IV) or triamminotrichloroplatinum(1+)

b. diamminodiaquadichlorocobalt(III) or diamminodiaquadichlorocobalt(1+)

c. diamminodiaquabromochlorocobalt(III) or diamminodiaquabromochlorocobalt(1+)

d. triaquabromochloroiodochromium(III) or triaquabromochloroiodochromium(0)

e. dichlorobis(ethylenediamine)platinum(IV) or dichlorobis(ethylenediamine)platinum(2+) or dichlorobis(1,2-ethanediamine)platinum(IV) or dichlorobis(1,2-ethanediamine)platinum(2+)

f. diamminodichloro(*o*-phenanthroline)chromium(III) or diamminodichloro(*o*-phenanthroline)chromium(1+) or diamminodichloro(1,10-phenanthrene)chromium(III) or diamminodichloro(1,10-phenanthrene)chromium(1+)

g. bis(bipyridine)bromochloroplatinum(IV) or bis(bipyridine)bromochloroplatinum(2+) or bis(2,2'-bipyridyl)bromochloroplatinum(IV) or bis(2,2'-bipyridyl)bromochloroplatinum(2+)

h. tris(methyldithiocarbamato)iron(III) or tris(methyldithiocarbamato)iron(0) or tris(methylcarbamyldithioato)iron(III) or tris(methylcarbamyldithioato)iron(0)

i. dibromo[*o*-phenylene(dimethylarsine)(dimethylphosphine)]rhenium(II) or dibromo[*o*-phenylene(dimethylarsine)(dimethylphosphine)]rhenium(0) or dibromo[1,2-phenylene(dimethylarsine)(dimethylphosphine)]rhenium(II) or dibromo[1,2-phenylene(dimethylarsine)(dimethylphosphine)]rhenium(0)

j. dibromochlorodiethylenetriaminerhenium(III) or dibromochlorodiethylenetriaminerhenium(0) or dibromochloro(2,2'-diaminodiethylamine)rhenium(III) or dibromochloro(2,2'-diaminodiethylamine)rhenium(0)

9-11 **a.** *cis*-dicarbonylbis(dimethyldithiocarbamato)ruthenium (III) or *cis*-dicarbonylbis(dimethyldithiocarbamato)ruthenium (0)

b. trisoxalatocobaltate(III) or trisoxalatocobaltate(3-)

c. tris(ethylenediamine)ruthenium(II) or tris(ethylenediamine)ruthenium(2+)

d. bis(2,2′-bipyridine)dichloronickel(II) or bis(2,2′-bipyridine)dichloronickel(2+)

9-12 **a.** tetraamminecopper(II) or tetraamminecopper(2+)

b. tetrachloroplatinate(II) or tetrachloroplatinate(2−)

c. tris(dimethyldithiocarbamato)iron(III) or tris(dimethyldithiocarbamato)iron(0)

d. hexacyanomanganate(II) or hexacyanomanganate(4-)

e. nonahydridorhenate(VII) or nonahydridorhenate(2-)

f. diamminesilver(I) tetrafluoroborate(III) or diamminesilver(1+) tetrafluoroborate(1-)

g. dicyanobis(methylisocyano)iron(II) or dicyanobis(methylisocyano)iron(0)

h. carbonatobis(ethylenediamine)cobalt(III) bromide or carbonatobis(ethylenediamine) cobalt(1+) bromide

i. pentaammineazidocobalt(III) sulfate or pentaammineazidocobalt(2+) sulfate

9-13. a. diamminebromochloroplatinum(II)

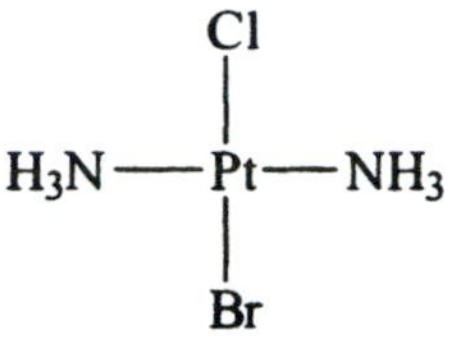

trans

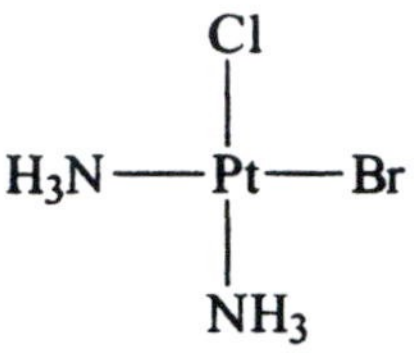

cis

b. diaquadiiododinitritopalladium(IV)

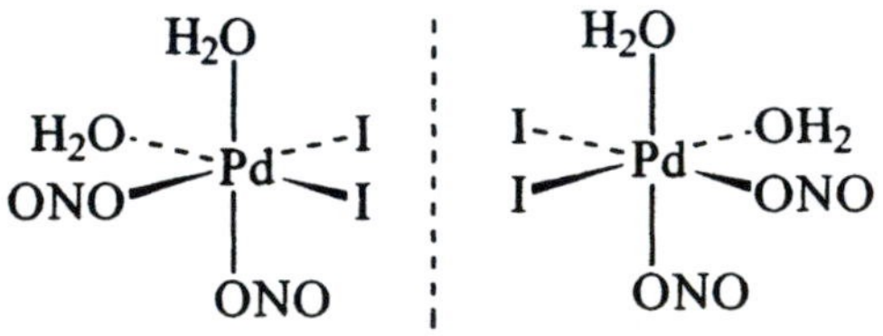

cis enantiomers

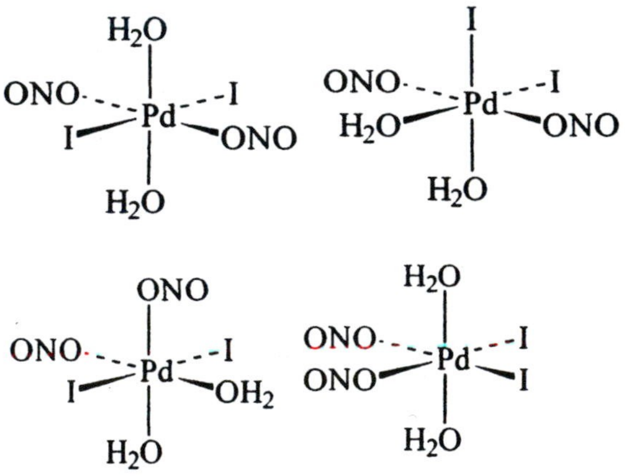

c. tri-μ-carbonylbis(tricarbonyliron(0))

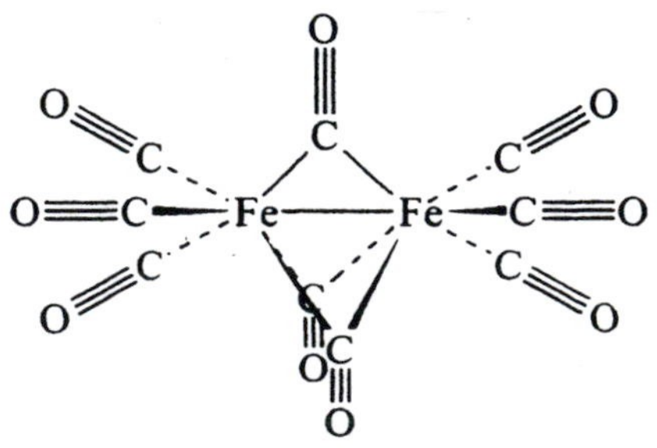

9-14 There are 18 isomers overall, twelve with the chelating ligand in a *mer* geometry and six with the chelating ligand in a *fac* geometry. All are enantiomers. They are all shown below, with dashed lines separating the enantiomers.

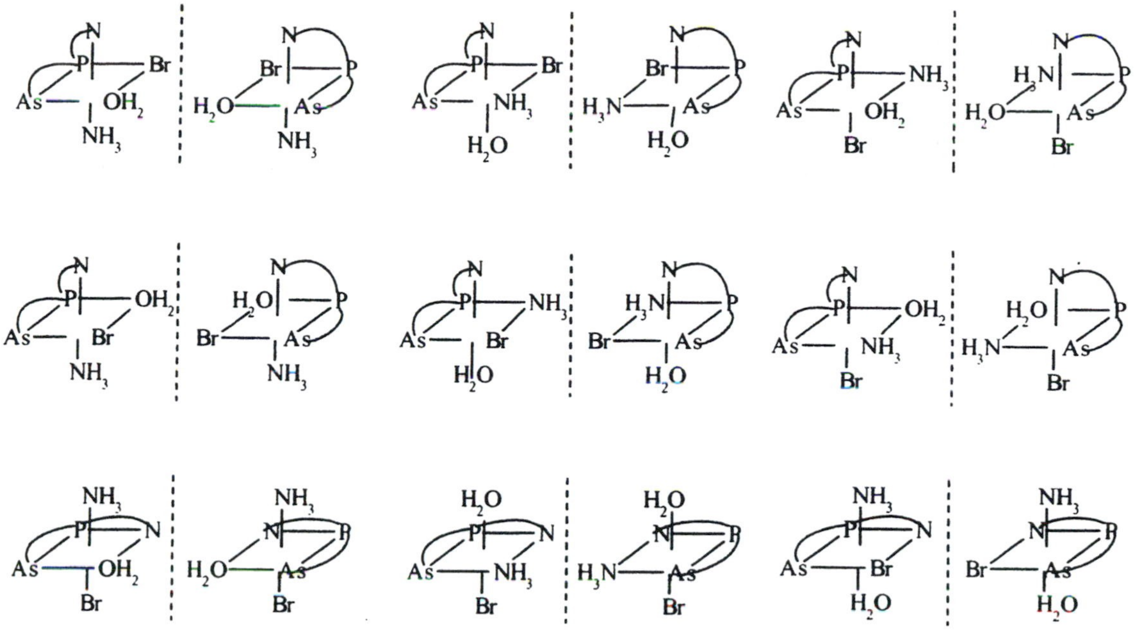

9-15 **a.** Δ **b.** Δ **c.** Λ **d.** Δ

9-16 All are chiral if the ring in **b** does not switch conformations.

9-17 15b δ 15c top ring:δ, bottom ring: λ

9-18

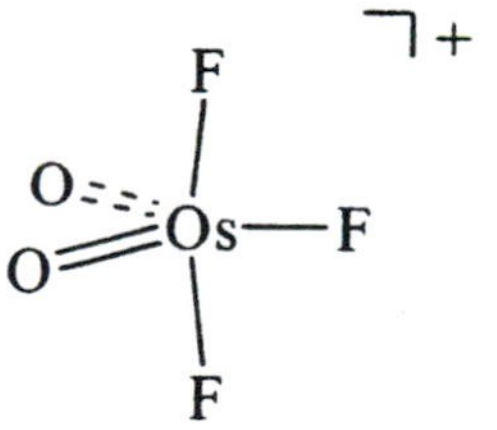

C_{2v} point group

The ^{19}F doublet is from the two axial fluorines (split by the equatorial fluorine).

The ^{19}F triplet is from the equatorial fluorines (split by the two aaxial fluorines).

The two doubly bonded oxygens are equatorial, as expected from VSEPR considerations.

9-19 Examples include both cations and anions:

$[Cu(CN)_2]^-$, $[Cu_2(CN)_3]^-$, $[Cu_3(CN)_4]^-$, $[Cu_4(CN)_5]^-$, $[Cu_5(CN)_6]^-$

$[Cu_2(CN)]^+$, $[Cu_3(CN)_2]^+$, $[Cu_4(CN)_3]^+$, $[Cu_5(CN)_4]^+$, $[Cu_6(CN)_5]^+$

Based primarily on calculations (rather than experimental data), Dance et al. have proposed linear structures such as the following:

$[Cu(CN)_2]^-$:	NC—Cu—CN
$[Cu_2(CN)_3]^-$:	NC—Cu—CN—Cu—CN
$[Cu_3(CN)_4]^-$:	NC—Cu—CN—Cu—CN—Cu—CN
$[Cu_2(CN)]+$	Cu—CN—Cu
$[Cu_3(CN)_2]+$	Cu—NC—Cu—CN—Cu
$[Cu_4(CN)_3]+$	Cu—CN—Cu—NC—Cu—CN—Cu

Where 2-coordinate copper appears in these ions, the geometry around the Cu is linear, as expected from VSEPR.

9-20 Since the phosphine ligand has C_3 symmetry and the two *ortho* methyl groups prevent free rotation about the P-C bonds, these ligands behave essentially as right- or left-handed propellers at low temperatures. There are four possible combinations, LL, LR, RR, and RL. In the solid phase, only RR and LL are found, with staggered C_2 configurations. In solution at low temperature (-60°C), two sets of ligand ^1H signals are seen. At temperatures above 30°C, these are reduced to three peaks for the CH_3 groups and two for the CH protons in the 3 and 5 positions as a result of rapid conversion from one form to the other.

Chapter 10
Coordination Chemistry II: Bonding

10-1 **a.** Tetrahedral d^6, 4 unpaired electrons

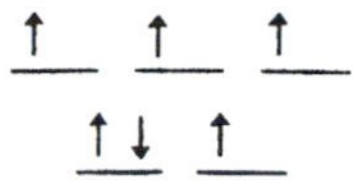

b. $[Co(H_2O)_6]^{2+}$, high spin octahedral d^7, 3 unpaired electrons

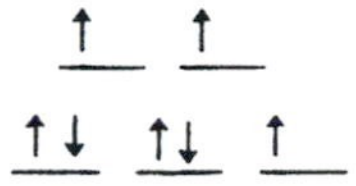

c. $[Cr(H_2O)_6]^{3+}$, octahedral d^3, 3 unpaired electrons

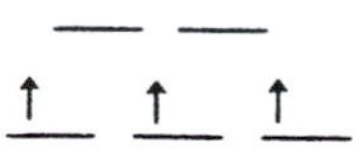

d. square planar d^7, 1 unpaired electron

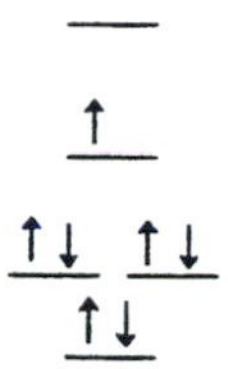

e. $5.1 \text{ BM} = \mu = \sqrt{n(n+2)}$; $n = 4.2 \approx 4$

10-2 $[Fe(CN)_6]^{4-}$ is Fe(II), low spin d^6, 0 unpaired electrons, diamagnetic

$[Co(H_2O)_6]^{3+}$ is Co(III), low spin d^6, 0 unpaired electrons, diamagnetic

$[CoF_6]^{3-}$ is Co(III), high spin d^6, 4 unpaired electrons (the only Co(III) high-spin compound)
$\mu = \sqrt{n(n+2)} = \sqrt{4\times 6} = \sqrt{24} = 4.9$ Bohr Magnetons

$[RhF_6]^{3-}$ is Rh(III), low spin d^6, 0 unpaired electrons, diamagnetic

10-3 $Fe(H_2O)_4(CN)_2$ is really $[Fe(H_2O)_6]_2[Fe(CN)_6]$, all Fe(II). $[Fe(H_2O)_6]^{2+}$ is high spin d^6, with $\mu = 4.9$; $[Fe(CN)_6]^{4-}$ is low spin d^6, with $\mu = 0$. The average value is then $2 \times 4.9/3 = 3.3$. 2.67 unpaired electrons gives $\mu = \sqrt{2.67 \times 4.67} = 3.53$.

10-4 Both $[M(H_2O)_6]^{2+}$ and $[M(NH_3)_6]^{2+}$ should show the double-humped curve of Figure 10-7, with larger values for the NH_3 compounds. Therefore, the difference between these curves still shows some of the same shape.

10-5 Simple σ angular overlap calculations for d^8 and d^9 ions show no energy difference between D_{4h} and O_h (Both d^8 geometries have energies of $-3\ e_\sigma$; both d^9 geometries have energies of $-6e_\sigma$). In general, stability constants decrease as more ligands are added, so the sequence for nickel is the common one. The huge drop in stability constant between the second and third ethylenediamine on Cu^{2+} is a result of the d^9 Jahn-Teller effect. The first two en molecules add in a square-planar geometry, with water molecules in the axial positions, and the difference between the ligands allows for the Jahn-Teller distortion. Adding a third en molecule requires a shift in geometry and pushes the bond distances closer together. This is counter to the Jahn-Teller distortion, and as a result the third addition is much less favorable than the first two.

10-6 All e_σ values are higher in energy, all e_π values are lower.

a. ML_2, using positions 1 and 6:

	e_σ	e_π	total
z^2	2	0	$2e_\sigma$
x^2-y^2	0	0	0
xy	0	0	0
xz	0	-2	$-2e_\pi$
yz	0	-2	$-2e_\pi$

b. ML_3, using positions 2, 11, 12:

	e_σ	e_π	total
z^2	0.75	0	$0.75e_\sigma$
x^2-y^2	1.125	-1.5	$1.125e_\sigma - 1.5e_\pi$
xy	1.125	1.5	$1.125e_\sigma - 1.5e_\pi$
xz	0	-1.5	$-1.5e_\pi$
yz	0	-1.5	$-1.5e_\pi$

c. ML_5, C_{4v}, using positions 1, 2, 3, 4, 5:

	e_σ	e_π	total
z^2	2	0	$2e_\sigma$
x^2-y^2	3	0	$3e_\sigma$
xy	0	-4	$-4e_\pi$
xz	0	-3	$-3e_\pi$
yz	0	-3	$-3e_\pi$

d. ML_5, D_{3h}, using positions 1, 2, 6, 11, 12:

	e_σ	e_π	total
z^2	2.75	0	$2.75e_\sigma$
x^2-y^2	1.125	-1.5	$1.125e_\sigma - 1.5e_\pi$
xy	1.125	-1.5	$1.125e_\sigma - 1.5e_\pi$
xz	0	-3.5	$-3.5e_\pi$
yz	0	-3.5	$-3.5e_\pi$

e. ML_8, cube, positions 7, 8, 9, 10, doubled for the other four corners:

	e_σ	e_π	total
z^2	0	-5.33	$-5.33e_\pi$
x^2-y^2	0	-5.33	$-5.33e_\pi$
xy	2.67	-1.78	$2.67e_\sigma - 1.78e_\pi$
xz	2.67	-1.78	$2.67e_\sigma - 1.78e_\pi$
yz	2.67	-1.78	$2.67e_\sigma - 1.78e_\pi$

10-7 **a.** The σ bonding energies are given in problem 10-6d.

b. For an axial L′, the π bonding would affect only the d_{xz} and d_{yz} orbitals, with energies stabilized by $1e_\pi$ for each. In the equatorial position 2, L′ lowers the energy of the xy and yz orbitals by $1e_\pi$ each. In positions 11 or 12, $d_{x^2-y^2}$ and d_{yz} are lowered by 0.75 e_π and d_{xy} and d_{xz} are lowered by 0.25 e_π. The average effect is that d_{z^2} is unaffected and the other four orbitals are stabilized by 0.50 e_π.

c. Stability depends on the number of *d* electrons; 1 through 7 electrons result in a stabilization of the axial isomer, 8 through 10 electrons result in no difference between the isomers.

10-8 The angular overlap differences (calculated in problem 10-6, c and d) between square pyramidal and trigonal bipyramidal structures are not large, but generally favor square pyramidal structures. Other factors (such as VSEPR) also contribute.

σ only	High spin		Low spin	
Number of *d* electrons	Square pyramidal	Trigonal bipyramidal	Square pyramidal	Trigonal bipyramidal
1	0	0	0	0
2	0	0	0	0
3	0	1.125 e_σ	0	0
4	2 e_σ	2.25 e_σ	0	0
5	5 e_σ	5 e_σ	0	1.125 e_σ
6	5 e_σ	5 e_σ	0	2.25 e_σ
7	5 e_σ	5 e_σ	2 e_σ	3.375 e_σ
8	5 e_σ	6.125 e_σ	5 e_σ	4.25 e_σ
9	7 e_σ	7.25 e_σ	7 e_σ	7.25 e_σ
10	10 e_σ	10 e_σ	10 e_σ	10 e_σ

σ and π	High spin		Low spin	
Number of *d* electrons	Square pyramidal	Trigonal bipyramidal	Square pyramidal	Trigonal bipyramidal
1	$-4\ e_\pi$	$-4\ e_\pi$	$-4\ e_\pi$	$-3.5\ e_\pi$
2	$-7\ e_\pi$	$-7\ e_\pi$	$-8\ e_\pi$	$-7\ e_\pi$
3	$-10\ e_\pi$	$1.125\ e_\sigma - 8.5\ e_\pi$	$-11\ e_\pi$	$-10.5\ e_\pi$
4	$2\ e_\sigma - 10\ e_\pi$	$2.25\ e_\sigma - 10\ e_\pi$	$-14\ e_\pi$	$-14\ e_\pi$
5	$5\ e_\sigma - 10\ e_\pi$	$5\ e_\sigma - 13.5\ e_\pi$	$-17\ e_\pi$	$1.125\ e_\sigma - 15.5\ e_\pi$
6	$5\ e_\sigma - 14\ e_\pi$	$5\ e_\sigma - 17\ e_\pi$	$-20\ e_\pi$	$2.25\ e_\sigma - 17\ e_\pi$
7	$5\ e_\sigma - 17\ e_\pi$	$5\ e_\sigma - 17\ e_\pi$	$2\ e_\sigma - 20\ e_\pi$	$3.375\ e_\sigma - 18.5\ e_\pi$
8	$5\ e_\sigma - 20\ e_\pi$	$6.125\ e_\sigma - 18.5\ e_\pi$	$5\ e_\sigma - 20\ e_\pi$	$4.25\ e_\sigma - 20\ e_\pi$
9	$7\ e_\sigma - 20\ e_\pi$	$7.25\ e_\sigma - 20\ e_\pi$	$7\ e_\sigma - 20\ e_\pi$	$7.25\ e_\sigma - 20\ e_\pi$
10	$10\ e_\sigma - 20\ e_\pi$	$10\ e_\sigma - 20\ e_\pi$	$10\ e_\sigma - 20\ e_\pi$	$10\ e_\sigma - 20\ e_\pi$

10-9 Co(II) is d^7. In tetrahedral complexes, it is high spin and has 3 unpaired electrons; in octahedral complexes, it is high spin and has 3 unpaired electrons; in square planar complexes, it has 1 unpaired electron. The magnetic moments can be calculated as $\mu = \sqrt{n(n+2)}$ = 3.9, 3.9, and 1.7, respectively.

10-10 For the red compounds (Me and Et at high temperatures, Pr, pip, pyr at all temperatures), the larger magnetic moment indicates approximately 5 unpaired electrons, appropriate for high-spin Fe(III) species. At low temperatures for the Me and Et compounds, the magnetic moment indicates 3 to 4 unpaired electrons, apparently an average value indicating an equilibrium mixture of high and low spin species. The low spin octahedral complexes would have one unpaired electron. Increasing the size of the R groups changes the structure enough that it is locked into the high-spin species at all temperatures.

10-11

		LFSE	Δ_o	LFSE	Difference
$[Co(H_2O)_6]^{3+}$	d^6 (ls)	$-2.4\Delta_o$	18,000	-43,200	35,440
$[Co(H_2O)_6]^{2+}$	d^7 (hs)	$-0.8\Delta_o$	9,700	-7,760	
$[Co(NH_3)_6]^{3+}$	d^6 (ls)	$-2.4\Delta_o$	24,000	-57,600	49,440
$[Co(NH_3)_6]^{2+}$	d^7 (hs)	$-0.8\Delta_o$	10,200	-8,160	

The ammine Co(III) complex is considerably more stable and is less easily reduced, with the difference primarily in the 3+ species. In addition, the metal in $[Co(NH_3)_6]^{3+}$ is surrounded by ammonia molecules, which are harder to oxidize than water. This makes transfer of electrons through the ligand more difficult for the ammine complex.

10-12

	n	μ	Term	LFSE (in Δ_o)
$[Co(CO)_4]^-$	0	0	^1S	0
$[Cr(CN)_6]^{4-}$	2	2.8	^3H	-1.6
$[Fe(H_2O)_6]^{3+}$	5	5.9	^6S	0
$[Co(NO_2)_6]^{4-}$	3	3.9	^4F	-0.8
$[Co(NH_3)_6]^{3+}$	0	0	^1I	-2.4
$[MnO_4]^-$	0	0	^1S	0
$[Cu(H_2O)_6]^{2+}$	1	1.7	^2D	-0.6

a. The two tetrahedral ions ($[Co(CO)_4]^-$ and MnO_4^-) have zero LFSE (10 or 0 *d* electrons, respectively) and can have π bonding, with CO as an acceptor and O^{2-} as donor. With the exception of $[Fe(H_2O)_6]^{3+}$, the others have LFSE values that favor octahedral structures. $[Fe(H_2O)_6]^{3+}$ has LFSE = 0 for either octahedral or tetrahedral shapes, but water is a slight π donor and Fe(III) is only a moderately good π acceptor. As a result, electrostatics favors six ligands.

b. High-spin octahedral d^7 has LFSE = $-0.8\Delta_o$
Tetrahedral d^7 has LFSE = $-1.2\Delta_t = -0.53\Delta_o$
High-spin octahedral d^8 has LFSE = $-1.2\Delta_o$
Tetrahedral d^8 has LFSE = $-0.6\Delta_t = -0.27\Delta_o$
Co(II) (d^7) has only $-0.27\Delta_o$ favoring the octahedral shape, while Ni(II) (d^8) has $-0.93\Delta_o$. Therefore, Co(II) compounds are more likely to be tetrahedral than are Ni(II) compounds.

10-13 Cl^- has the lowest Δ_o value and fairly good π donor properties that reduce Δ_o. F^- is next, with less π donor ability. Water has very small π donor ability (only one lone pair not involved in σ bonding) and ammonia and en have neither π donor nor acceptor ability (no lone pairs, antibonding orbitals with the wrong shapes and energies for bonding). CN^- has good π acceptor properties, making Δ_o largest for this ligand.

10-14 a. Compression moves d_{z^2} up in energy and lowers the ligand energies of positions 1 and 6.

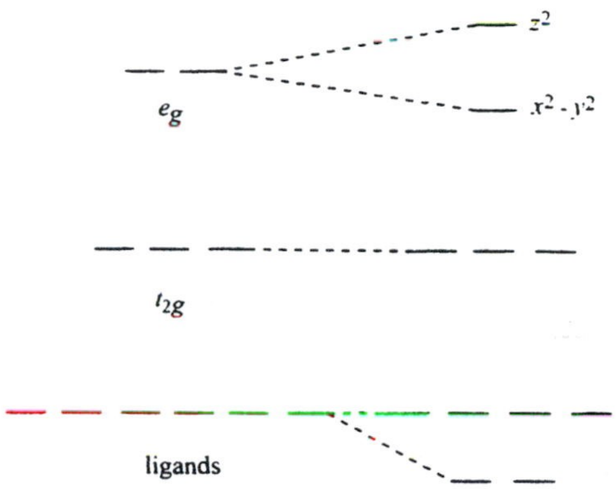

b. Stretching reverses the changes. In the limit of a square planar structure, d_{z^2} is affected only through interactions with the ring in the *xy* plane.

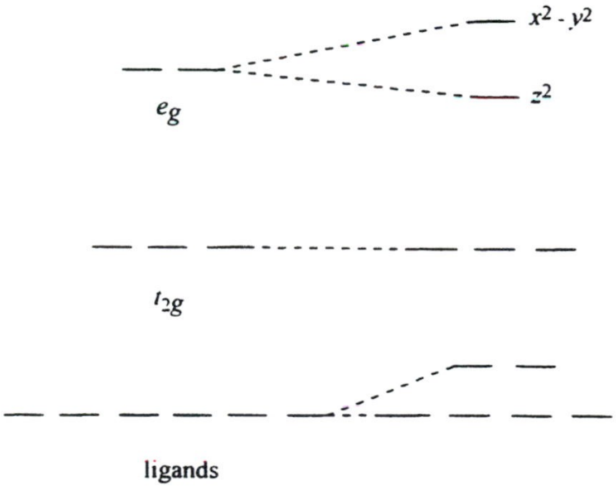

10-15 a.

	O_h LFSE	T_d LFSE	O_h- T_d
d^1	$-0.4\Delta_o$	$-0.6\Delta_t = -0.27\Delta_o$	$-0.13\Delta_o$
d^2	$-0.8\Delta_o$	$-1.2\Delta_t = -0.53\Delta_o$	$-0.27\Delta_o$
d^3	$-1.2\Delta_o$	$-0.8\Delta_t = -0.36\Delta_o$	$-0.84\Delta_o$
d^4	$-0.6\Delta_o$	$-0.4\Delta_t = -0.18\Delta_o$	$-0.42\Delta_o$
d^5	$0\Delta_o$	$0\Delta_t = 0\Delta_o$	$0\Delta_o$
d^6	$-0.4\Delta_o$	$-0.6\Delta_t = -0.27\Delta_o$	$-0.13\Delta_o$
d^7	$-0.8\Delta_o$	$-1.2\Delta_t = -0.53\Delta_o$	$-0.27\Delta_o$
d^8	$-1.2\Delta_o$	$-0.8\Delta_t = -0.36\Delta_o$	$-0.84\Delta_o$
d^9	$-0.6\Delta_o$	$-0.4\Delta_t = -0.18\Delta_o$	$-0.42\Delta_o$
d^{10}	$0\Delta_o$	$0\Delta_t = 0\Delta_o$	$0\Delta_o$

The differences in the *d* orbital energies always favor octahedral complexes, with the order of the absolute values $d^5 < d^6 < d^7 < d^9 < d^8$. This order predicts the relative stability of tetrahedral complexes as Mn > Fe > Co > Cu > Ni. Fe is out of order, Co and Ni are in the correct order.

b. Angular overlap calculations result in $-e_\sigma$ for each ligand electron, or $-12e_\sigma$ for an octahedral complex and $-8e_\sigma$ for a tetrahedral complex. The metal electrons have the energies shown in the table on the next page.

	O_h	T_d	O_h- T_d
d^1	$-12e_\sigma$	$-8e_\sigma$	$-4e_\sigma$
d^2	$-12e_\sigma$	$-8e_\sigma$	$-4e_\sigma$
d^3	$-12e_\sigma$	$-6.67e_\sigma$	$-5.33e_\sigma$
d^4	$-9e_\sigma$	$-5.33e_\sigma$	$-3.67e_\sigma$
d^5	$-6e_\sigma$	$-4e_\sigma$	$-2e$
d^6	$-6e_\sigma$	$-4e_\sigma$	$-2e_\sigma$
d^7	$-6e_\sigma$	$-4e_\sigma$	$-2e_\sigma$
d^8	$-6e_\sigma$	$-2.67e_\sigma$	$-3.33e_\sigma$
d^9	$-3e_\sigma$	$-1.33e_\sigma$	$-1.67e_\sigma$
d^{10}	$0e_\sigma$	$0e_\sigma$	$0e_\sigma$

Again, all cases favor octahedral geometry, but here the relative order for tetrahedral geometry is $d^9 > d^6 = d^7 > d^8$. Nickel and cobalt are in the correct order, but iron is equal to cobalt rather than less likely to have tetrahedral geometry.

10-16 The energy levels of the square planar *d* orbitals are shown in Figure 10-13 (p. 358); those of the octahedral orbitals are shown in Figure 10-5 (p. 347). The seventh, eighth, and ninth *d* electrons in an octahedral complex go into the highest orbitals, raising the total energy of the complex. In these cases, the energy of the square planar complex may be more favorable even though the total ligand electron energy is less than in octahedral complexes because there are only four ligands rather than six.

10-17 Ammonia is a stronger field ligand than water. It is a stronger Lewis base (σ donor) than water. In addition, water has a lone pair that can act as a π donor. These are unfavorable enough that the less electronegative nitrogen on ammonia is a better ligand. In the halide ions, all have the same electronic structure, so the electronegativity is the determining factor in ligand field strength. Fluoride is also a stronger base than the other halide ions.

10-18 Square pyramidal complexes have C_{4v} symmetry.

C_{4v}	E	$2C_4$	C_2	$2\sigma_v$	$2\sigma_d$	
Γ	5	1	1	3	1	
A_1	1	1	1	1	1	z, z^2
B_1	1	-1	1	1	-1	x^2-y^2
E	2	0	-2	0	0	(x,y) (xz,yz)

The z^2 and x^2-y^2 orbitals are the major *d* orbitals used in the bonding, so the *d* orbital energies are as shown in the figure. The z^2 orbital is less involved as it is directed at one ligand and the x^2-y^2 is directed at four ligands.

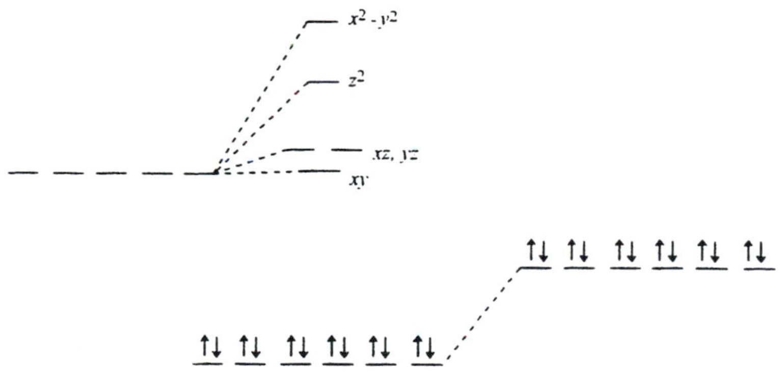

10-19 Cr^{3+} has three singly occupied t_{2g} orbitals and two empty e_g orbitals. As a result, there is no Jahn-Teller distortion in its complexes. Mn^{3+} has one electron in the e_g orbitals, so there is Jahn-Teller distortion in its complexes.

10-20 MnO_4^- has no *d* electrons, while MnO_4^{2-} has one. The slight antibonding effect of this electron is enough to lengthen the bonds.

10-21 **a.** $[Cr(H_2O)_6]^{2+}$ $n = 4$

$\mu = \sqrt{4(6)} = 4.9$ BM

b. $[Cr(CN)_6]^{4-}$ $n = 2$

$\mu = \sqrt{2(4)} = 2.8$ BM

c. $[FeCl_4]^-$ $n = 5$

$\mu = \sqrt{5(7)} = 5.9$ BM

d. $[Fe(CN)_6]^{3-}$ $n = 1$

$\mu = \sqrt{1(3)} = 1.7$ BM

e. $[Ni(H_2O)_6]^{2+}$ $n = 2$

$\mu = \sqrt{2(4)} = 2.8$ BM

f. $[Cu(en)_2(H_2O)_2]^{2+}$ $n = 1$

$\mu = \sqrt{1(3)} = 1.7$ BM

10-22 Metal *d* orbitals, NH_3 influence:

	e_σ	e_π	total
z^2	1	0	$1e_\sigma$
x^2-y^2	3	0	$3e_\sigma$
xy	0	0	0
xz	0	0	0
yz	0	0	0

Metal *d* orbitals, Cl^- influence

	e_σ	e_π	total
z^2	2	0	$2e_\sigma$
x^2-y^2	0	0	0
xy	0	0	0
xz	0	2	$2e_\pi$
yz	0	2	$2e_\pi$

Ligand NH_3:

	e_σ	total
1	0	0
2	-1	$-1e_\sigma$
3	-1	$-1e_\sigma$
4	-1	$-1e_\sigma$
5	-1	$-1e_\sigma$
6	0	0

Ligand Cl^-:

	e_σ	e_π	total
1	-1	-2	$-1e_\sigma - 2e_\pi$
2	0	0	0
3	0	0	0
4	0	0	0
5	0	0	0
6	-1	-2	$-1e_\sigma - 2e_\pi$

Overall energy = $-8e_\sigma(NH_3) - 4e_\sigma(Cl) - 8e_\pi(Cl) + 4e_\pi(Cl) = -8e_\sigma(NH_3) - 4e_\sigma(Cl) - 4e_\pi(Cl)$
The metal electrons are still unpaired, one in the undisturbed d_{xy} orbital and one each in d_{xz} and d_{yz}, raised by π interaction with Cl^-. Four of the ligand orbitals are lowered by $e_\sigma(NH_3)$ and two are lowered by $e_\sigma(Cl) + 2e_\pi(Cl)$; each contains a pair of ligand electrons. Since the magnitudes of the changes are unknown, the order of ligand orbital energies is uncertain.

10-23 a. The new positions (13 and 14 below) are opposite 11 and 12, and have the same values in the table. z^2 is affected most strongly, since two ligands are along the z axis; $x^2 - y^2$ and xy are also strongly influenced, since they are in the plane of the six ligands. xz and yz are not changed, since they miss the ligands in all directions.

	z^2	$x^2 - y^2$	xy	xz	yz
1	1	0	0	0	0
2	1/4	3/4	0	0	0
6	1	0	0	0	0
11	1/4	3/16	9/16	0	0
12	1/4	3/16	9/16	0	0
4	1/4	3/4	0	0	0
13	1/4	3/16	9/16	0	0
14	1/4	3/16	9/16	0	0
Sum	3.5	2 1/4	2 1/4	0	0

b. From the table below, we can see that $z^2 = A_{1g}$, $(x^2 - y^2, xy) = E_{2g}$ and $(xz, yz) = E_{1g}$

D_{6h}	E	$2C_6$	$2C_3$	C_2	$3C_2'$	$3C_2''$	i	$2S_3$	$2S_6$	σ_h	$3\sigma_d$	$3\sigma_v$		
Γ	8	2	2	2	2	0	0	0	0	6	4	2		
A_{1g}	1	1	1	1	1	1	1	1	1	1	1	1		z^2
A_{2g}	1	1	1	1	-1	-1	1	1	1	1	-1	-1		
B_{1g}	1	-1	1	-1	1	-1	1	-1	1	-1	1	-1		
B_{2g}	1	-1	1	-1	-1	1	1	-1	1	-1	-1	1		
E_{1g}	2	1	-1	-2	0	0	2	1	-1	-2	0	0		(xz, yz)
E_{2g}	2	-1	-1	2	0	0	2	-1	-1	-2	0	0		$(x^2 - y^2, xy)$
A_{1u}	1	1	1	1	1	1	-1	-1	-1	-1	-1	-1		
A_{2u}	1	1	1	1	-1	-1	-1	-1	-1	-1	1	1	z	
B_{1u}	1	-1	1	-1	1	-1	-1	1	-1	1	-1	1		
B_{2u}	1	-1	1	-1	-1	1	-1	1	-1	1	1	-1		
E_{1u}	2	1	-1	-2	0	0	-2	-1	1	2	0	0	(x, y)	
E_{2u}	2	-1	-1	2	0	0	-2	1	1	-2	0	0		

$\Gamma = 2A_{1g}$	+	E_{2g}	+	A_{2u}	+	B_{2u}	+	E_{1u}
s, z^2		$(x^2 - y_2, xy)$		z				(x, y)

c.

	z^2	$x^2 - y^2$	xy	xz	yz
1	0	0	0	1	1
2	0	0	1	1	0
6	0	0	0	1	1
11	0	3/4	1/4	1/4	3/4
12	0	3/4	1/4	1/4	3/4
4	0	0	1	1	0
13	0	3/4	1/4	1/4	3/4
14	0	3/4	1/4	1/4	3/4
Sum	0	$3\ e_\pi$	$3\ e_\pi$	$5\ e_\pi$	$5\ e_\pi$
Overall	$3.5\ e_\sigma$	$2.5\ e_\sigma + 3\ e_\pi$	$2.5\ e_\sigma + 3\ e_\pi$	$5\ e_\pi$	$5\ e_\pi$

$3.5\ e_\sigma$ ——

$2.5\ e_\sigma + 3\ e_\pi$ —— ——

Uncoordinated Metal _ _ _ _ _

$5\ e_\pi$ —— ——

Because the Jahn-Teller effect is larger when the lowest level is unevenly populated, d^1 and d^3 would have the largest effect. The location of $d_{x^2-y^2}$ and d_{xy} relative to the uncoordinated metal ion is uncertain; it depends on the magnitudes of e_σ and e_π.

Chapter 11
Coordination Chemistry III: Electronic Spectra

11-1 **a.** p^3 There will be (6!)/(3!3!) = 20 microstates

		M_S			
		$-3/2$	$-1/2$	$1/2$	$3/2$
	$+2$		$1^+\ 1^-\ 0^-$	$1^+\ 1^-\ 0^+$	
	$+1$		$1^+\ 1^-\ -1^-$ $1^-\ 0^-\ 0^+$	$1^+\ 1^-\ -1^+$ $1^+\ 0^-\ 0^+$	
M_L	0	$1^-\ 0^-\ -1^-$	$1^+\ 0^-\ -1^-$ $1^-\ 0^+\ -1^-$ $1^-\ 0^-\ -1^+$	$1^+\ 0^+\ -1^-$ $1^+\ 0^-\ -1^+$ $1^-\ 0^+\ -1^+$	$1^+\ 0^+\ -1^+$
	-1		$-1^+\ -1^-\ 1^-$ $-1^-\ 0^-\ 0^+$	$-1^+\ -1^-\ 1^+$ $-1^+\ 0^-\ 0^+$	
	-2		$-1^+\ -1^-\ 0^-$	$-1^+\ -1^-\ 0^+$	

Terms: $L = 0, S = 3/2$: ^4S (ground state) $L = 2, S = 1/2$: ^2D $L = 1, S = 1/2$: ^2P

b. p^1d^1 There will be (6!)/(1!5!) (10!)/(1!9!) = 60 microstates

		M_S		
		-1	0	1
	3	$1^-\ 2^-$	$1^-\ 2^+$, $1^+\ 2^-$	$1^+\ 2^+$
	2	$1^-\ 1^-$ $0^-\ 2^-$	$1^+\ 1^-$, $1^-\ 1^+$ $0^-\ 2^+$, $0^+\ 2^-$	$1^+\ 1^+$ $0^+\ 2^+$
	1	$1^-\ 0^-$ $0^-\ 1^-$ $-1^-\ 2^-$	$1^+\ 0^-$, $0^+\ 1^-$ $1^-\ 0^+$, $0^-\ 1^+$ $-1^-\ 2^+$, $-1^+\ 2^-$	$1^+\ 0^+$ $0^+\ 1^+$ $-1^+\ 2^+$
M_L	0	$1^-\ -1^-$ $0^-\ 0^-$ $-1^-\ 1^-$	$1^+\ -1^-$, $1^-\ -1^+$ $0^-\ 0^+$, $0^+\ 0^-$ $-1^-\ 1^-$, $-1^-\ 1^+$	$1^+\ -1^+$ $0^+\ 0^+$ $-1^+\ 1^+$
	-1	$-1^-\ 0^-$ $0^-\ -1^-$ $1^-\ -2^-$	$-1^+\ 0^-$, $0^+\ -1^-$ $-1^-\ 0^+$, $0^-\ -1^+$ $1^-\ -2^+$, $1^+\ -2^-$	$-1^+\ 0^+$ $0^+\ -1^+$ $1^+\ -2^+$
	-2	$-1^-\ -1^-$ $0^-\ -2^-$	$-1^+\ -1^-$, $-1^-\ -1^+$ $0^-\ -2^+$, $0^+\ -2^-$	$-1^+\ -1^+$ $0^+\ -2^+$
	-3	$-1^-\ -2^-$	$-1^-\ -2^+$, $-1^+\ -2^-$	$-1^+\ -2^+$

Terms: $L = 3, S = 1$ ^3F (ground state) $L = 2, S = 0$ ^1D
$L = 3, S = 0$ ^1F $L = 1, S = 1$ ^3P
$L = 2, S = 1$ ^3D $L = 1, S = 0$ ^1P

The two electrons are independent of each other, because they are in different orbitals. Both can have the same m_l and m_s values, because they have different l values.

11-2 For p^3, $L = 0$, $S = 3/2$, the term ^4S has $J = 3/2$ only ($|L+S| = |L-S|$). Therefore, the ground state is $^4S_{3/2}$.

For p^1d^1, $L = 3$, $S = 1$, the term ^3F has $J = 4, 3, 2$. Since both levels are less than half-filled, the lowest J has the lowest energy, and the ground state is $^3F_2$.

11-3 **a.** ^2D has $L = 2$ and $S = 1/2$, so $M_L = -2, -1, 0, 1, 2$ and $M_S = -1/2, 1/2$

b. ^3G has $L = 4$ and $S = 1$, so $M_L = -4, -3, -2, -1, 0, 1, 2, 3, 4$ and $M_S = -1, 0, 1$

c. ^4F has $L = 3$ and $S = 3/2$, so $M_L = -3, -2, -1, 0, 1, 2, 3$ and $M_S = -3/2, -1/2, 1/2, 3/2$

11-4 **a.** ^2D with $J = 5/2, 3/2$ fits an excited state of d^5, $^2D_{3/2}$

b. ^3G with $J = 5, 4, 3$ fits an excited state of d^4, $^3G_3$

c. ^4F with $J = 9/2, 7/2, 5/2, 3/2$ fits the ground state of d^7, $^4F_{9/2}$

11-5

$$\varepsilon = 0.038\frac{L}{mol\ cm} \quad A = 0.10 \quad l = 1.00cm$$

$$A = \varepsilon lc; \quad c = \frac{A}{\varepsilon\, l} = \frac{0.10}{(0.038\frac{L}{mol\ cm})(1.00cm)} = 2.6\ \frac{mol}{L}$$

11-6 ***a.***

$$\bar{\nu} = 24{,}900\ cm^{-1}$$

$$\lambda = \frac{1}{\bar{\nu}} = \frac{1}{24{,}900\ cm^{-1}} = 4.02\times10^{-5}\ cm = 402\ nm$$

$$\nu = \frac{c}{\lambda} = \frac{(2.998\times10^8\ m\ s^{-1})(100\ cm\ m^{-1})}{(4.02\times10^{-5}\ cm)} = 7.47\times10^{14}s^{-1}$$

b.

$$\lambda = 366\ nm$$

$$\nu = \frac{c}{\lambda} = \frac{(2.998\times10^8\ m\ s^{-1})(10^9\ nm\ m^{-1})}{366\ nm} = 8.19\times10^{14}\ s^{-1}$$

$$E = h\nu = (6.626\times10^{-34}\ J\ s)(8.19\times10^{14}\ s^{-1}) = 5.43\times10^{-19}\ J$$

$$\bar{\nu} = \frac{1}{\lambda} = \frac{1}{366\ nm}(10^7\ nm\ cm^{-1}) = 27{,}300\ cm^{-1}$$

11-7. J values are included in these answers although they are not essential for the spectral analysis.

a. d^8 O_h $M_S = 1 = S$ Spin multiplicity = 2+1 = 3
Max M_L = 2+2+1+1+0+0-1-2 = 3 = L $J = |L \pm S| = 4, 3, 2$ $^3F_4$

b. d^5 O_h high spin $M_S = 5/2 = S$ Spin multiplicity = 5+1 = 6
Max M_L = 2+1+0-1-2 = 0 = L $J = |L \pm S| = 5/2$ $^6S_{5/2}$

d^5 O_h low spin $M_S = 1/2 = S$ Spin multiplicity = 1+1 = 2
Max M_L = 2+2+1+1+0 = 6 = L $J = |L \pm S|$ = 11/2, 13/2 $^2I_{11/2}$, $^2I_{13/2}$
(J is uncertain in this case; the usual rule does not apply because the level is exactly half full.)

c. d^4 T_d $M_S = 2 = S$ Spin multiplicity = 4+1 = 5
Max M_L = 2+1+0-1 = 2 = L $J = |L \pm S|$ = 4, 3, 2, 1, 0 $^5D_0$

d. d^9 D_{4h} $M_S = 1/2 = S$ Spin multiplicity = 1+1 = 2
Max M_L = 2+2+1+1+0+0-1-1-2 = 2 = L $J = |L \pm S|$ = 5/2, 3/2 $^2D_{5/2}$

11-8 $[Ni(H_2O)_6]^{2+}$ For d^8 ions, the the energy of the lowest energy band is Δ_o, so Δ_o = 8,700 cm^{-1}. The bands are split due to Jahn-Teller distortion in the excited state.

11-9 a. $[Cr(C_2O_4)_3]^{3-}$ is Cr(III), d^3. Δ_o is equal to the lowest energy band, so Δ_o = 17,400 cm^{-1}.

b. $[Ti(NCS)^6]^{3-}$ is Ti(III), d^1. Δ_o is the energy of the single band; Δ_o = 18,400 cm^{-1}. The band is split due to Jahn-Teller distortion of the excited state.

c. $[Ni(en)_3]^{2+}$ is Ni(II), d^8. The lowest energy band corresponds to Δ_o; Δ_o = 11,200 cm^{-1}

d. $[VF_6]^{3-}$ is V(III), d^2. Following the example on pp. 402-405, we find the ratio ν_2/ν_1 and then Δ_o/B: ν_2/ν_1 = 1.57 at Δ_o/B = 26. From the Tanabe-Sugano diagram at Δ_o = 26,
ν_1: E/B = 24.1 E = 24.1 B = 14,800 cm^{-1} B = 614 cm^{-1}
ν_2: E/B = 37.0 E = 37.0 B = 23,250 cm^{-1} B = 628 cm^{-1}
Average B = 620 cm^{-1}, Δ_o = 26 B = 16,000 cm^{-1}

e. V(III) is a d^2 ion. Again following the example on pp. 402-405, ν_2 = 21,413 cm^{-1} and ν_1 = 14,409 cm^{-1}, ν_2/ν_1 = 1.49. From the Tanabe-Sugano diagram at Δ_o = 34.5,
ν_1: E/B = 29 E = 29 B = 18,100 cm^{-1} B = 489 cm^{-1}
ν_2: E/B = 44 E = 44 B = 18,100 cm^{-1} B = 489 cm^{-1}
Average B = 489 cm^{-1}, Δ_o = 34.5 B = 16,870 cm^{-1}

11-10. $[Co(NH_3)_6]^{2+}$, d^7. As in problem 11-9d:
ν_2/ν_1 = 2.34 at Δ_o/B = 11. From the Tanabe-Sugano diagram at Δ_o = 11,
ν_1: E/B = 10 E = 10 B = 9,000 cm^{-1} B = 900 cm^{-1}
ν_2: E/B = 22.5 E = 22.5 B = 21,100 cm^{-1} B = 938 cm^{-1}
Average B = 920 cm^{-1}, Δ_o = 11 B = 10,100 cm^{-1}

11-11 **a.** $t_{2g}^{4}e_{g}^{2}$ The t_{2g} level is a triply degenerate asymmetrically occupied state, so it is *T*.

b. t_{2g}^{6} This is a nondegenerate state, completely occupied, so it is *A*.

c. $t_{2g}^{3}e_{g}^{3}$ This is an excited state, with the t_{2g} level uniformly occupied and the e_g level doubly degenerate, so it is *E*.

d. t_{2g}^{5} Vacancies behave the same as electrons, so this is a triply degenerate state, *T*.

e. e_g Another excited state, this is doubly degenerate, *E*.

11-12 The complexes with potential degeneracies are those with d^1, d^2, d^4, d^7, and d^9, low spin d^5, and high spin d^6 configurations. The strongest effects are with high spin d^4, low spin d^7, and d^9 complexes, corresponding to $[Mn(NH_3)_6]^{3+}$, $[Ni(NH_3)_6]^{3+}$, and $[Cu(NH_3)_6]^{3+}$. Weaker effects might be seen with $[Ti(NH_3)_6]^{3+}$, $[V(NH_3)_6]^{3+}$, low spin $[Fe(NH_3)_6]^{3+}$, and high spin $[Co(NH_3)_6]^{3+}$, all of which are unknown or unstable. Other ligands are needed to stabilize ions containing these 3+ metal ions.

11-13 Rhenium is significantly larger than manganese and the 5*d* orbitals of Re are higher in energy than the 3*d* orbitals of Mn, so a CTTM excitation requires more energy for ReO_4^-. In addition, since the molecular orbitals derived primarily from the 3*d* orbitals of MnO_4^- are lower in energy than the corresponding MO's of ReO_4^-, MnO_4^- is better able to accept electrons; it is a better oxidizing agent.

11-14 The order of energy of the charge transfer bands is I < Br < Cl. The HOMO of the iodide complex is highest in energy, and the LUMO's are similar in energy, so the difference is smallest for iodide. In other terms, iodide is softer and can lose an electron most easily in a CTTM process.

11-15 Comparing $[Fe(CN)_6]^{3-}$ (low spin d^5) and $[Fe(CN)_6]^{4-}$ (low spin d^6), where CN^- is a σ donor and a π acceptor: The t_{2g} orbitals of $[Fe(CN)_6]^{3-}$ contain 5 electrons, allowing CTTM from the ligand orbitals to either t_{2g} or e_g levels. The t_{2g} levels of $[Fe(CN)_6]^{4-}$ are full, so only the higher e_g levels are available for CTTM. CTTL transitions ($t_{2g} \longrightarrow \pi^*$) excitations are also possible for either complex.

11-16 **a.** O^{2-} and Cl^- are both σ and π donors, and the metal ions are Cr(V) and Mo(V) (the ligands are Cl^- and O^{2-}). Metal *d* orbitals, Cl^- influence:

	e_σ	e_π	total
z^2	2	0	$2e_\sigma$
x^2-y^2	3	2	$3e_\sigma + 2e_\pi$
xy	0	2	$2e_\pi$
xz	0	2	$2e_\pi$
yz	0	2	$2e_\pi$

Metal d orbitals, O^{2-} influence:

	e_σ	e_π	total
z^2	1	0	e_σ
x^2-y^2	0	0	0
xy	0	0	0
xz	0	1	e_π
yz	0	1	e_π

Overall, from bottom to top, the d orbitals are

xy $(+2e_\pi(Cl))$

xz and yz $(+2e_\pi(Cl)+e_\pi(O))$

z^2 $(+2e_\sigma(Cl)+e_\sigma(O))$

x^2-y^2 $(3e_\sigma(Cl)+2e_\pi(Cl)$

b. These complexes have C_{4v} symmetry. The symmetry of the orbitals is given in the character table:

C_{4v}	E	$2C_4$	C_2	$2\sigma_v$	$2\sigma_d$	
A_1	1	1	1	1	1	z^2
B_1	1	-1	1	1	-1	x^2-y^2
B_2	1	-1	1	-1	1	xy
E	2	0	-2	0	0	(xz, yz)

Each of the representations is labeled with the matching orbital(s).

c. The lowest d orbital has an energy of $2e_\pi(Cl)$; the next have energies of $2e_\pi(Cl)$ + $e_\pi(O)$. As d^1 complexes, the transition is between these orbitals; the transition is from an essentially nonbonding orbital to a π* orbital. The interaction between metal and ligand are generally stronger for a second row transition metal than for the first row, raising the π* energy in the Mo case.

11-17 $V(CO)_6]^-$ < $Cr(CO)_6$ < $[Mn(CO)_6]^+$ As the nuclear charge on the metal increases, the metal orbitals are drawn down to lower energies. Consequently, the CTTL bands should increase in energy.

11-18 a. At 80K: $\mu_s = 0.65 = \sqrt{n(n+2)}$ $n = 0.19$

At 300K: $\mu_s = 5.2 = \sqrt{n(n+2)}$ $n = 4.3$

b. The complex is near the low spin-high spin boundary of the d^6 Tanabe-Sugano diagram. High spin becomes increasingly favored as the temperature increases.

11-19 a. ML_2, using positions 1 and 6, with O^{2-} both a σ and π donor:

	e_σ	e_π	total
z^2	2	0	$2e_\sigma$
x^2-y^2	0	0	0
xy	0	0	0
xz	0	2	$2e_\pi$
yz	0	2	$2e_\pi$

b. If this is a high spin complex, there are four electrons in the lowest levels (xy, x^2-y^2), three in the next two (xz, yz), and one in the highest (z^2). Electronic transitions can be either from the middle levels to the top, and from the bottom levels to the middle and the top, three possibilities in all.

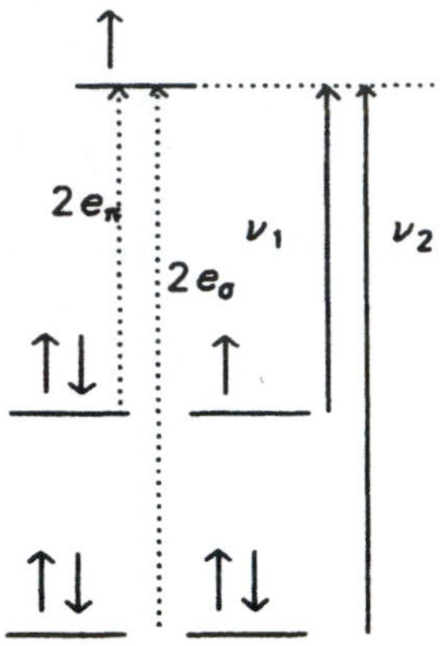

According to Hitchman, Stratemeier, and Hoppe, the transitions seen are from the middle and the bottom levels to the top level.

c. Assigning the transitions as in **b**: $E = 2e_\sigma = 16{,}000\text{cm}^{-1}$ $E = 2e_\pi = 9{,}000$ cm^{-1} ,
so $e_\sigma = 8{,}000\ \text{cm}^{-1}$ and $e_\pi = 4{,}500\ \text{cm}^{-1}$

11-20 $Re(CO)_3(P(OPh_3))(DBSQ)$ 18,250 cm^{-1}

$Re(CO)_3(PPh_3)(DBSQ)$ 17,300 cm^{-1}

$Re(CO)_3(NEt_3)(DBSQ)$ 16,670 cm^{-1}

NEt_3 is the strongest donor ligand in this series. Therefore, the metal in the complex $Re(CO)_3(NEt_3)(DBSQ)$ has the greatest concentration of electrons on the metal and the greatest tendency to transfer electrons to acceptor orbitals. Since this complex also has the lowest energy charge transfer band, we may assign this as charge transfer to ligand.

11-21 a. RuO_4^{2-} has the highest value of Δ_t. Δ_t increases with the oxidation state of the metal and in general is greater for second row than for first row metals. The overall trend is $RuO_4^{2-} > FeO_4^{2-} > MnO_4^{3-} > CrO_4^{4-}$.

b. The nuclear charge of iron is greatest in this isoelectronic series and exerts the strongest attraction for bonding electrons. As a result, FeO_4^{2-} has the shortest metal-oxygen distance.

c. As the nuclear charge of the metal increases, the metal orbitals are pulled in to lower energies. Consequently, less energy is needed to excite electrons from ligand orbitals to metal orbitals. These are ligand to metal charge transfers.

11-22 Addition of aqueous NH_3 replaces the H_2O ligands in the green $[Ni(H_2O)_6]^{2+}$ to give blue $[Ni(NH_3)_6]^{2+}$. As a bidentate ligand, en can replace two NH_3 ligends; three en ligands can replace all six NH_3 ligands to form violet $[Ni(en)_3]^{2+}$:

	$[Ni(H_2O)_6]^{2+}$	$\xrightarrow{NH_3}$	$[Ni(NH_3)_6]^{2+}$	$\xrightarrow{en}$	$[Ni(en)_3]^{2+}$
	green		blue		violet
Complementary color:	red		orange		yellow

The complementary colors in this series have increasing energies, indicating that en has the strongest effect on Δ_o, H_2O the weakest. This is consistent with the positions of these ligands in the spectrochemical series.

11-23 a. These colors are most likely the consequence of CTTM transisions, from orbitals that are primarily from the oxide ligands to orbitals that are primarily from the metal:

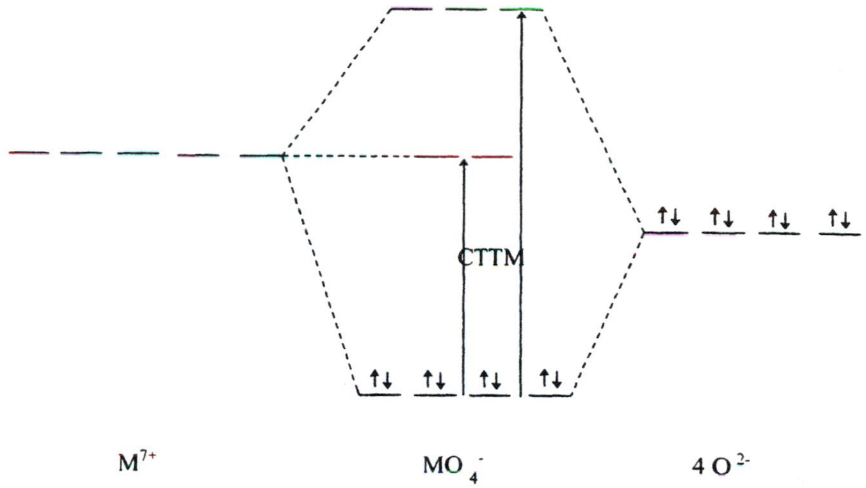

b. In TcO_4^-, the separation between the donor orbitals of the O^{2-} ligands and the acceptor orbitals is greater than in MnO_4^-. As a consequence, TcO_4^- absorbs light of higher energy (green) than MnO_4^- (yellow). Actually, most of the absorption by TcO_4^- is in the untraviolet, with the pale red color a result of a low-intensity tail of the absorption band extending into the visible.

c. The metal-ligand interactions in MnO_4^{2-} (Mn(VI)) is weaker than in MnO_4^- (Mn(VII)), and the separation of donor and acceptor orbitals in MnO_4^{2-} is smaller, meaning that less energy (red light) is necessary for excitation than in MnO_4^- (yellow). Also worth noting: the Mn—O bond distance is longer in MnO_4^{2-} (165.9 pm) than in MnO_4^- (162.9 pm), an indication of weaker bonding in the former.

11-24 a. At 350 nm:

$$\varepsilon = \frac{A}{lc} = \frac{2.34}{(1.00\text{ cm})(2.00 \times 10^{-4}\text{ M})} = 11{,}700\text{ L mol}^{-1}\text{ cm}^{-1}$$

At 514 nm:

$$\varepsilon = \frac{0.532}{(1.00\text{ cm})(2.00 \times 10^{-4}\text{ M})} = 2{,}640\text{ L mol}^{-1}\text{ cm}^{-1}$$

At 590 nm:

$$\varepsilon = \frac{0.370}{(1.00 \text{ cm})(2.00 \times 10^{-4} \text{ M})} = 1{,}850 \text{ L mol}^{-1} \text{ cm}^{-1}$$

At 1540 nm:

$$\varepsilon = \frac{0.0016}{(1.00 \text{ cm})(2.00 \times 10^{-4} \text{ M})} = 8.0 \text{ L mol}^{-1} \text{ cm}^{-1}$$

b. Because of their high intensity, the bands at 350, 514, and 590 nm are probably charge transfer bands. However, the low molar absorptivity of the band at 1540 nm indicates that it is probably a *d-d* transition (see examples in Figure 11-8).

11-25 a. From Problem 11-1a, for a p^3 configuration there are three terms, ^4S, ^2D, and ^2P. The J values for each of these are determined below.

For ^4S: $L = 0$, $S = \frac{3}{2}$; Because $J = L + S, L + S - 1, \ldots|L - S|$, the quantum number J can only be $\frac{3}{2}$ and there is a single state for ^4S: $^4S_{3/2}$

For ^2D: $L = 2$, $S = \frac{1}{2}$. Possible J values are $\frac{5}{2}$ and $\frac{3}{2}$, and the two possible states are $^2D_{5/2}$ and $^2D_{3/2}$.

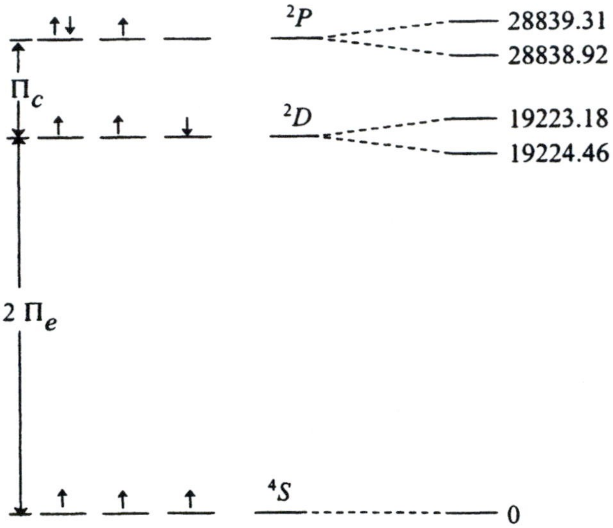

For ^2P: $L = 1$, $S = \frac{1}{2}$. Possible J values are $\frac{3}{2}$ and $\frac{1}{2}$, with states $^2P_{3/2}$ and $^2P_{1/2}$.

The lowest energy state is $^4S_{3/2}$ (highest multiplicity). $^2D_{5/2}$ and $^2D_{3/2}$ are next, at 19233.18 and 19224.46 cm^{-1}, and $^2P_{3/2}$ and $^2P_{1/2}$ are the highest energy at 28828.92 and 28839.31 cm^{-1}.

b. The difference in energy between the ^4S and ^2D states is $2\Pi_e$. From the average of the two nearly degenerate ^2D states, $\Pi_e = 9614.41$ cm^{-1}.

The difference in energy between the averages of the ^2D and ^2P states is $\Pi_c = 28{,}839.12 - 19{,}228.82 = 9610.30$ cm^{-1}.

11-26 These are all d^8 complexes, with three excited states of the same spin multiplicity as the ground state. For d^8, Δ_o = energy of the lowest energy band. B can be calculated by using the method of Problem 11-9. Δ_o is the difference between the transitions $^3A_2 \rightarrow {}^3T_1$ and $^3A_2 \rightarrow {}^3T_2$; a graph of ν_2/ν_1 versus Δ_o/B needs to be prepared for d^8 Ni^{2+} in order to calculate B, as described in pages 404-405. A plot of the ratio of the highest energy band to the lowest energy band is shown here.

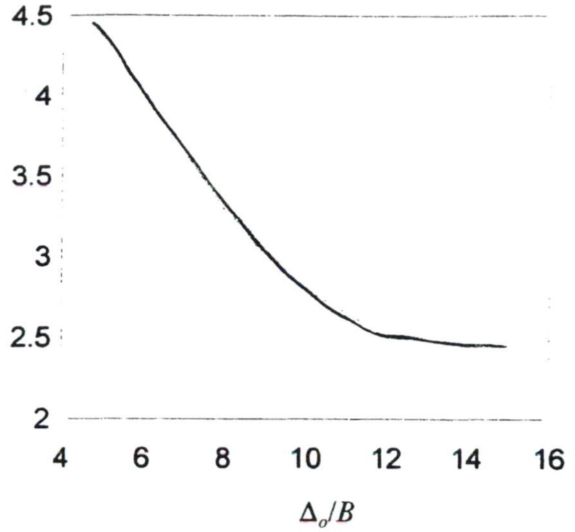

Species	Ratio	Δ_o/B	Δ_o (cm^{-1})	B (cm^{-1})
$[Ni(H_2O)_6]^{2+}$	3.06	8.8	8,500	970
$[Ni(NH_3)_6]^{2+}$	2.62	12.5	10,500	860
$[Ni(OS(CH_3)_2)_6]^{2+}$	3.11	8.4	7,728	920
$[Ni(dma)_6]^{2+}$	3.14	8.0	7,576	950

11-27 a. These are both high spin d^7 complexes, for which the ground state term symbol is ^4F (see Figure 11-7).

b. There are three possible transitions, all originating from the $^4T_1(^4F)$ and going to the $^4T_2$, $^4A_2$ and $^4T_1(^4P)$ levels.

c. For $[Co(bipy)_3]^{2+}$, the ratio $\frac{\nu_2}{\nu_1} = \frac{22000\ cm^{-1}}{11300\ cm^{-1}} = 1.95$. From Figure 11-14, this comes at $\frac{\Delta_o}{B} = 17$, at which:

for the 22000 cm^{-1} band: $\frac{E}{B} = 28$ and $B = \frac{22000\ cm^{-1}}{28} = 786\ cm^{-1}$

for the 11,300 cm^{-1} band: $\frac{E}{B} = 15$ and $B = \frac{11300\ cm^{-1}}{15} = 754\ cm^{-1}$

Average $B = 770\ cm^{-1}$; $\Delta_o = 17\ B = 13{,}100\ cm^{-1}$

For a d^7 complex, LFSE = $-\frac{4}{5}\Delta_o = -10{,}500\ cm^{-1}$

Δ_o for $[Co(NH_3)_6]^{2+}$ was calculated in Problem 11-10; $\Delta_o = 10{,}100\ cm^{-1}$.

For a d^7 complex, LFSE = $-\frac{4}{5}\Delta_o = -8{,}080 \text{ cm}^{-1}$

d. These bands should be broad (see Figure 11-8), a consequence of vibrational motion within these complexes.

e. The molecular orbital energy level diagrams should be similar to Figure 10-5, with 6 electrons in t_{2g} orbitals and one electron in an e_g level in each case. The separation between the t_{2g} and e_g orbitals should be larger in the bipy complex.

Chapter 12
Coordination Chemistry IV: Reactions and Mechanisms

12-1 $[Cr(H_2O)_6]^{2+}$ is labile, and has 4 unpaired electrons, with one in the e_g level. This electron in an antibonding orbital makes substitution relatively easy. $[Cr(CN)_6]^{4-}$ is inert. It has all 4 electrons in the t_{2g} levels. Substitution requires activation into the antibonding e_g level, so reactions are slow. The strong π bonding with CN^- also strengthens the Cr-C bonds.

12-2 The kinetic equations for I_a and I_d mechanisms are similar, so use of the rate equation to distinguish between them is not possible. However, if the rate constants for different entering ligands are quite different, it is clear that increasing the coordination number is important in the rate determining step. If dissociation is more important, the rates should be different for different leaving groups.

12-3 Pentachlorooxochromate(V) is a d^1 complex; it should be labile, with vacancies in the t_{2g} levels.

Hexaiodomanganate(IV) is a d^3 complex; it should be inert, much like Cr(III).

Hexacyanoferrate(III) is a low spin d^5 complex; it should be inert (vacant e_g levels).

Hexammineiron(III) is a d^5 high spin complex; with partly occupied orbitals in both levels, it should be labile.

12-4 $[Fe(H_2O)_6]^{3+}$ and $[Co(H_2O)_6]^{2+}$ are high-spin species, and the electrons in the upper e_g levels make them labile. $[Cr(CN)_6]^{4-}$ is a d^4 low spin species. The t_{2g} levels are unequally occupied and the e_g are vacant, which makes it a borderline complex in terms of rate. $[Cr(CN)_6]^{3-}$, $[Fe(CN)_6]^{4-}$, and $[Co(NH_3)_5(H_2O)]^{3+}$ are all low-spin species with the t_{2g} levels either half filled or completely filled. This, combined with empty e_g levels, indicates inert species. LFAE numbers indicate that activation of these ions requires large amounts of energy.

12-5 The general rate law for square planar substitution is
Rate = $(k_1 + k_2[Cl^-])\,[Pt(NH_3)_4^{2+}]$.
The general procedure is to measure either the disappearance of $[Pt(NH_3)_4]^{2+}$ or the formation of $[Pt(NH_3)_3Cl]^+$ to find the rate. The most convenient method is by UV or visible absorbance spectra, using a wavelength where there is a large difference in absorbance between the two species.

a. Swamp the solution with chloride, so that its concentration does not change. This is likely to make the reaction appear first order in Pt reactant. Measure the change in concentrations, plot ln $[Pt(NH_3)_4^{2+}]$ vs. time to get k_{obs}.

b. Repeat several times with different chloride concentrations.

c. Graph k_{obs} vs. $[Cl^-]$. If the rate law above is followed, the result is a straight line with intercept = k_1 and slope = k_2.
If the equation above does not hold, the first order approximation in a will not work, and the log plot will not give a straight line.

12-6 **a.** Exchange with DMSO is extremely fast, regardless of the X^- concentration, because DMSO is present in very large excess as the solvent. The rate constant is also larger than for the other reactants.

b. The general equation for this reaction is probably
Rate = k [cplx] [X]/(1 + k'[X]).
At low [X], the reactions are first order in [X] (1 >> k'[X]). At high [X], where 1 << k'[X], the reactions become zero order in [X].

c. The limiting rate constant is equal to k_2k_1 for D, and k_2K_1 for I_d.

d. Since the limiting rate constants are so similar, the major rate determining step is dissociative. However, the slight difference with differing X means that association has some part in the rate, and it is then I_d.

12-7 $[Cu(H_2O)_6]^{2+}$ is a d^9 complex, subject to Jahn-Teller distortion. Therefore, different rates are observed for exchange of axial and equatorial water molecules.

12-8 A plot of $[As(C_6H_5)_3]$ vs k_{obs} is a straight line, with intercept $k_1 = 2.3 \times 10^{-5}\ s^{-1}$ and slope $k_2 = 2.06 \times 10^{-5}\ M^{-1}s^{-1}$.

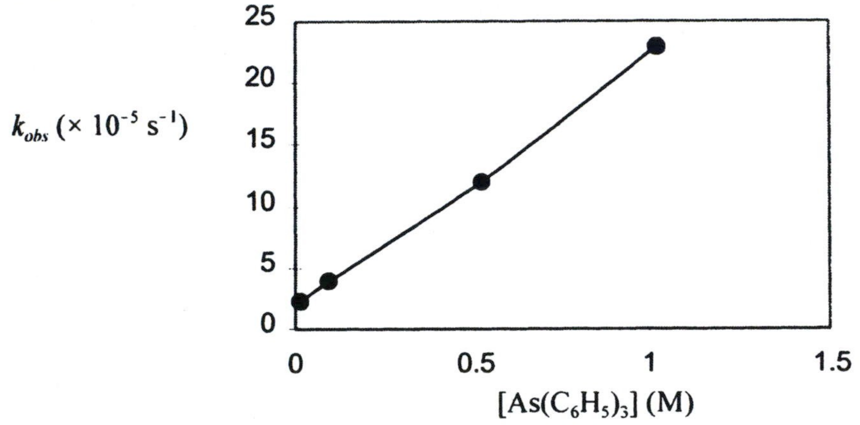

Rate = $(k_1 + k_2[As(C_6H_5)_3])[Co(NO)CO)_3]$

This reaction, like the one in problem 5, shows both first and second order kinetics. The first order reaction appears to be a dissociative reaction or a solvent-assisted dissociation of CO, followed by a fast addition of $As(C_6H_5)_3$. The other path shows first order dependence on $As(C_6H_5)_3$, probably caused by an associative reaction.

12-9 ΔHNP is a measure of basicity, with more basic molecules having a smaller ΔHNP. The reactions appear to be associative, with the rate increasing with the basicity of the incoming ligand. The two lines are a consequence of the different natures of the ligands; numbers 1-11 are all phosphorus ligands, while 12-14 are organic nitrogen compounds. The half neutralization potential provides a relative measure of basicity within compounds of similar structure, but is not an absolute measure of the reactivity of the compounds.

12-10 a. $\Delta G° = \Delta H° - T\Delta S° = 10300 - 298 \times 55.6 = -6300$ J/mol = -6.3 kJ/mol.
$\Delta G° = -RT \ln K$; $\ln K = -\Delta G°/RT = 6300/(8.3145 \times 298.15) = 2.53$; $K = 12.5$.

b. The *cis* isomer has the higher bond energy (actually, the lower overall energy), since rearrangement to the *trans* isomer requires addition of energy. Since the phosphines are better π acceptors, the *cis* isomer should be the more stable. If the phosphines are *trans* to each other, they compete for the electron density of the same *d* orbital electrons. In the *cis* position, each can use one of the pair of d_{xz} and d_{yz} orbitals and avoid competition.

c. The free phosphine must help the reaction by an associative mechanism. Since benzene is the solvent, and it is very nonpolar, it is not likely to assist the reaction. Only the second term of the rate equation of problem 12-5 is significant here, with phosphine playing the role of the entering ligand.

12-11 Two factors need to be considered:
1. The best π acceptors (CO, PPh_3) slow the dissociation of CO in the *cis* position; π donors (halides) speed the dissociation.
2. Those with very slow rates are d^6 species; the faster reacting complexes are d^5 species. All are low spin because of the large number of CO's. There are fewer electrons available for π back-bonding in the halide complexes, resulting in weaker Cr-CO bonds, in spite of the π donation by the halides.

12-12 *Trans*-$Pt(NH_3)_2Cl_2$ reacts with thiourea to form *trans*-$[Pt(NH_3)_2(tu)_2]^{2+}$; the first Cl^- is easily removed (*trans* effect and inherent lability) and the strong *trans* effect of tu then helps replace the second Cl^-. Both chlorides of *cis*-$Pt(NH_3)_2Cl_2$ are easily replaced (inherent lability of Cl). The tu's are then *trans* to the NH_3's, which are replaced because of the strong *trans* effect of tu, resulting in $[Pt(tu)_4]^{2+}$.

12-13 a. $[Pt(CO)Cl_3]^- + NH_3 \rightarrow$ *trans*-$[Pt(CO)(NH_3)Cl_2]$ CO is the stronger *trans* director

b. $[Pt(NH_3)Br_3]^- + NH_3 \rightarrow$ *cis*-$[Pt(NH_3)_2Br_2]$ Br is the stronger *trans* director

c. $[Pt(C_2H_4)Cl_3]^- + NH_3 \rightarrow$ *trans*-$[Pt(C_2H_4)(NH_3)Cl_2]$ C_2H_4 is the stronger *trans* director.

12-14 a. Two sets of reactions, with examples from Figure 12-12 identified:
$[PtCl_4]^{2-} + 2\,NH_3 \rightarrow$ *cis*-$[PtCl_2(NH_3)_2] + 2\,Cl^-$ (b)
cis-$[PtCl_2(NH_3)_2] + 2$ py $\rightarrow$ *cis*-$[Pt(py)_2(NH_3)_2]^{2+} + 2\,Cl^-$ (h)
cis-$[Pt(py)_2(NH_3)_2]^{2+} + 2\,NO_2^- \rightarrow$ *trans*-$[Pt(NO_2)_2(NH_3)(py)] + NH_3 +$ py (e)
trans-$[Pt(NO_2)_2(NH_3)(py)] + CH_3NH_2 \rightarrow$
$[Pt<(NO_2)(CH_3NH_2)><(NH_3)(py)>]^+ + NO_2^-$ (g)

and

$[PtCl_4]^{2-} + 2$ py $\rightarrow$ *cis*-$[PtCl_2(py)_2] + 2\,Cl^-$ (b)
cis-$[PtCl_2(py)_2] + 2\,CH_3NH_2 \rightarrow$ *cis*-$[Pt(CH_3NH_2)_2(py)_2]^{2+} + 2\,Cl^-$ (h)
cis-$[Pt(CH_3NH_2)_2(py)_2]^{2+} + 2\,NO_2^- \rightarrow$
trans-$[Pt(NO_2)_2(CH_3NH_2)(py)]$ + py + CH_3NH_2 (c)

$$trans\text{-}[Pt(NO_2)_2(CH_3NH_2)(py)] + NH_3 \rightarrow$$
$$[Pt<(NO_2)(NH_3)><(CH_3NH_2)(py)>]^+ + NO_2^- \qquad \text{(a)}$$

b. Reaction with Cl_2 puts Cl both above and below the plane, with the products

$[Pt<(NO_2)(NH_3)><(CH_3NH_2)(py)><(Cl)(Cl)>]^+$
and
$[Pt<(NO_2)(CH_3NH_2)><(NH_3)(py)><(Cl)(Cl)>]^+$

Reaction with one mole of Br^- replaces one Cl^-, with the products

$[Pt<(NO_2)(NH_3)><(CH_3NH_2)(py)><(Br)(Cl)>]^+$
and
$[Pt<(NO_2)(CH_3NH_2)><(NH_3)(py)><(Br)(Cl)>]^+$

12-15 V^{2+} is d^3, and likely to be inert. V^{3+} is d^2, and labile. In low $[H^+]$ (or slightly basic) solutions, V^{3+} becomes VOH^{2+}, which may substitute into V^{2+} more readily. The V-OH-V combination must be significant in the rate, with *a* and *b* determining the relative rates of the two paths. Both are probably inner sphere, but we have no direct evidence, just similarity to Cr^{2+}-Cr^{3+} reactions. The rate equation is

$$\text{Rate} = (k_1 + k_2/[H^+])[V^{2+}][V^{3+}], \text{ so } a = k_1\,[V^{2+}][V^{3+}] \text{ and } b = k_2[V^{2+}][V^{3+}]$$

12-16 $[Cr(H_2O)_6]^{2+}$ is labile, but $[Co(NH_3)_6]^{3+}$ is inert, and the NH_3's cannot bridge readily (no way for them to bond with the Cr). Therefore, the reaction is likely to be an outer-sphere reaction.

12-17 This reaction is inner sphere, so the X^- acts as a bridging ligand. The order of rates is $Br^- > Cl^- > N_3^- > F^- > NCS^-$. The rate may be based partly on the substitution rates on Cr^{2+}, but that is likely to be a small effect. Most of the difference is the difference in transmission of the electron through the ligand and transfer of the ligand to the other Cr. Br^- works best, since it is soft, but not too soft. F^- is too hard, and holds the electrons too well; NCS^- has the soft S end available for bonding to Cr^{2+}, which is hard. Based on structure, azide should be even faster, since it contains hard N and multiple bonds that usually transfer electrons readily. Perhaps the multiple bonds soften it, but this is only a lower limit on the rate. The data in this problem are from D. L. Ball and E. L. King, *J. Am. Chem. Soc.*, **1958**, *80*, 1091.

Chapter 13
Organometallic Chemistry

13-1 A: = method A, B: = method B

a.	$Fe(CO)_5$	A: 8 + 5×2 = 18	B: 8 + 5×2 = 18
b.	$[Rh(bipy)_2Cl]^+$	A: 7 + 2×4 + 2 = 17	B: 9 + 2×4 + 1 - 1 = 17
c.	$(\eta^5\text{-}Cp^*)Re(=O)_3$	A: 6 + 0 + 3×4 = 18	B: 5 + 7 + 3×2 = 18
d.	$Re(PPh_3)_2Cl_2N$	A: 2 + 2×2 + 2×2 + 6 = 16	B: 7 + 2×2 + 2×1 + 3 = 16
e.	$Os(CO)(\equiv CPh)(PPh_3)_2Cl$	A: 7 + 2 + 3 + 2×2 + 2 = 18	B: 8 + 2 + 3 + 2×2 + 1 = 18

a, **c**, and **e** obey the 18-electron rule.

13-2 All of these compounds have 16-electron valence configurations.

a.	$Ir(CO)Cl(PPh_3)_2$	A: 8 + 2 + 2 + 2×2 = 16	B: 9 + 2 + 1 + 2×2 = 16
b.	$RhCl(PPh_3)_3$	A: 8 + 2 + 3×2 = 16	B: 9 + 1 + 3×2 = 16
c.	$[Ni(CN)_4]^{2-}$	A: 8 + 4×2 = 16	B: 10 + 4×1 + 2 = 16
d.	*cis*-$PtCl_2(NH_3)_2$	A: 8 + 2×2 + 2×2 = 16	B: 10 + 2×2 + 2×1 = 16

13-3

a. $[M(CO)_7]^+$ — A: 18 - 7×2 = 4 = M^+, V — B: 18 - 7×2 = 4 = M^+, V

b. $H_3CM(CO)_5$ — A: 18 - 2 - 5×2 = 6 = M^+, Mn — B: 18 - 1 - 5×2 = 7 = M, Mn

c. $M(CO)_2(CS)(PPh_3)Br$ — A: 18 - 2×2 - 2 - 2 - 2 = 8 = M^+, Co
B: 18 - 2×2 - 2 - 2 - 1 = 9 =M, Co

d. $[\eta^3\text{-}C_3H_3)(\eta^5\text{-}C_5H_5)M(CO)]^-$ — A: 18 - 4 - 6 - 2 = 6 = M^+, Mn
B: 18 - 3 - 5 - 2 = 8 = M^-, Mn

e. $(OC)_5M{=}C(OCH_3)C_6H_5$ — A and B: 18 - 5×2 - 2 = 6 = M, Cr

f. $[\eta^4\text{-}C_4H_4)(\eta^5\text{-}C_5H_5)M]^+$ — A: 18 - 4 - 6 = 8 = M^{2+}, Ni — B: 18 - 4 - 5 = 9 = M^+, Ni

g. $(\eta^3\text{-}C_3H_5)(\eta^5\text{-}C_5H_5)M(CH_3)(NO)$

linear NO: A: 18 - 2 - 6 - 2 - 2 = 6 = M, Cr
B: 18 - 3 - 5 - 1 - 3 = 6 = M, Cr

bent NO: A: 18 - 2- 6 - 2 - 2= 6= M^{2+}, Fe
B: 18 - 3 - 5 - 1 - 1 = 8 = M, Fe

h. $[M(CO)_4I(diphos)]^-$ — A: 18 - 4×2 - 2 - 4 = 4 = M, Ti
B: 18 - 4×2 - 1 - 4 = 5 = M^-, Ti

13-4 Calculating for each metal atom:

a. $[Fe(CO)_2(\eta^5\text{-}C_5H_5)]_2$
A: 7+ 2×2 + 6= 17, single Fe-Fe
B: 8 + 2×2 + 5 = 17, single Fe-Fe

b. $[Mo(CO)_2(\eta^5\text{-}Cp)]_2^{2-}$
A: 5 + 2×2 + 6 + 1 = 16, double Mo=Mo
B: 6 + 2×2 + 5 + 1 = 16, double Mo=Mo

13-5 a. $[M(CO)_3(NO)]^-$
linear M–N–O:
A: 18 – 3×2 – 2 – 2 = 8 = M, Ru
B: 18 – 3×2 – 3 = 9 = M^-, Ru
bent M–N–O:
A: 18 – 3×2 – 2 = 10 = M, Pd
B: 18 – 3×2 – 1 = 11 = M^-, Pd

b. $[M(PF_3)_2(NO)_2]^+$
linear NO:
A: 18 – 2×2 – 2×2 = 10 = M^-, Rh
B: 18 – 2×2 – 2×3 = 8 = M^+, Rh

c. $[M(CO)_4(\mu_2\text{-}H)]_3$
As a triangular structure with three M–M bonds:
A: 18 – 4×2 – 2 – 2 = 6 = M^+, Tc
B: 18 – 4×2 – 1 – 2 = 7 = M, Tc

d. $M(CO)(PMe_3)_2Cl$
A: 16 – 2 – 2×2 – 2 = 8 = M^+, Rh
B: 16 – 2 – 2×2 – 1 = 9 = M, Rh

13-6 Method B works better for calculating overall charge.

a. $[Co(CO)_3]^z$ 9 + 3×2 = 15, z = 3–

b. $[Ni(CO)_3(NO)]^z$ 10 + 3×2 + 3 = 19, z = 1+

c. $[Ru(CO)_4(GeMe_3)]^z$ 8 + 4×2 + 1 = 17, z = 1–

d. $[(\eta^3\text{-}C_3H_5)V(CNCH_3)_5]^z$ 3 + 5 + 5×2 = 18, z = 0

e. $[(\eta^5\text{-}C_5H_5)Fe(CO)_3]^z$ 5 + 8 + 3×2 = 19, z = 1+

f. $[(\eta^5\text{-}C_5H_5)_3Ni_3(\mu_3\text{-}CO)_2]^z$ 3×5 + 3×10 + 2×2 = 49, z = 1+, assuming three Ni-Ni bonds; calculating for each Ni: 5 + 10 + 2(2/3) + 2 = 18 1/3; charge per Ni = 1/3+, overall charge = 1+

13-7 a. $[\eta^5\text{-}C_5H_5)W(CO)_x]_2$, assuming a single W–W:
A: 6 + 5 + 1 + x×2 = 18, x = 3
B: 5 + 6 + 1 + x×2 = 18, x = 3

b. $ReBr(CO)_x(CO_2C_2H_4)$
A: 6 + 2 + x×2 + 2 = 18, x = 4
B: 7 + 1 + x×2 + 2 = 18, x = 4

c. $[(CO)_3Ni\text{-}Co(CO)_3]^z$ A and B: $3\times2 + 10 + 2 + 9 + 3\times2 - z = 36$; $z = 3-$

d. $[Ni(NO)_3(SiMe_3)]^z$ B: $10 + 3\times3 + 1 - z = 18$, $z = 2+$

e. $[(\eta^5\text{-}C_5H_5)Mn(CO)_x]_2$ B: $5 + 7 + 2 + x\times2 = 18$, $x = 2$

13-8 Figure 10-18 (p. 361) gives an MO diagram for $Ni(CO)_4$ and other references. The HOMO's (t_2) are strongly bonding, and there is a large energy gap between the HOMO's and the LUMO's in this 18-electron molecule.

13-9 The energy of stretching vibrations depends on the square root of the force constant divided by the reduced mass (Section 13-4-1).

The reduced masses are 14.73 for ^{16}O and 16.41 for ^{18}O, so $(14.73/16.41)^{1/2} = 0.947$, and the ^{18}O complex has a vibrational energy of $0.947 \times 975\ cm^{-1} = 924\ cm^{-1}$. The value given in the reference is $926\ cm^{-1}$.

13-10 Sulfur is less electronegative than oxygen. Therefore, the tungsten in $W(S)Cl_2(CO)(PMePh_2)_2$ has greater electron density and a greater tendency to back-donate to CO; a lower energy $\nu(CO)$ is expected (actual value: $1986\ cm^{-1}$).

13-11 Adding electrons to a carbonyl complex puts more electrons into the back-bonding orbitals. As a result, the V–C bonds in $[V(CO)_6]^-$ are strengthened and the distance shortened (but the C–O bonds are weakened by having more electrons in the antibonding orbitals).

13-12

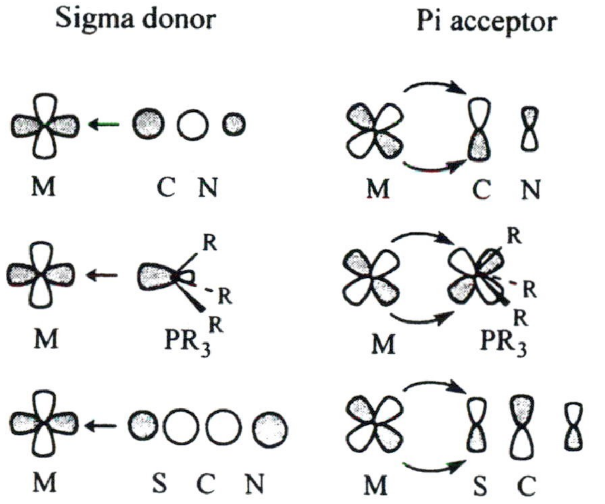

13-13 a. If NO is counted as a linear donor, each of these has 18 electrons. The increasing nuclear charge draws the NO electrons in toward the metal, resulting in less backbonding to the ligands and an increase in NO bond order. The Cr species has an energy below that of most linear species, so it may be a bent NO, making it a 16 electron ion. The Fe complex has an energy above the normal, so its π back-bonding must be unusually weak.

b. The low energy band indicates a bent NO ligand, with the higher energy band from the linear ligand. This complex has one of each, with angles of 138° and 178° (Greenwood and Earnshaw, *Chemistry of the Elements*, 2nd ed., pp. 450-52).

13-14 a. The CO_2 molecular orbitals are shown in Chapter 5, page147.

b. The π orbitals of 1,3,5-hexatriene are shown in Figure 13-21 on page 480.

c. *Cyclo*-C_4H_4

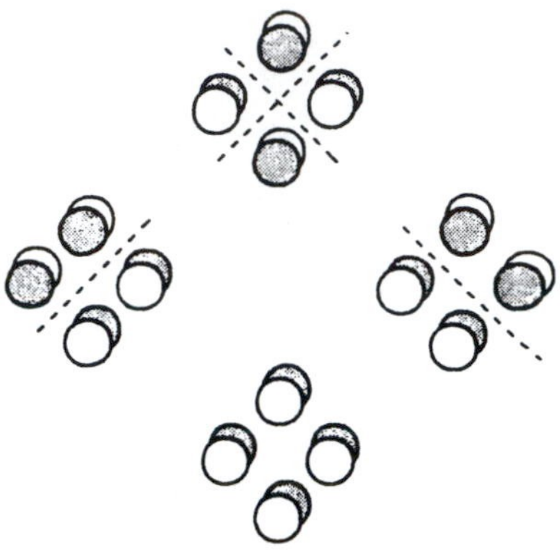

d. *Cyclo*-C_7H_7:

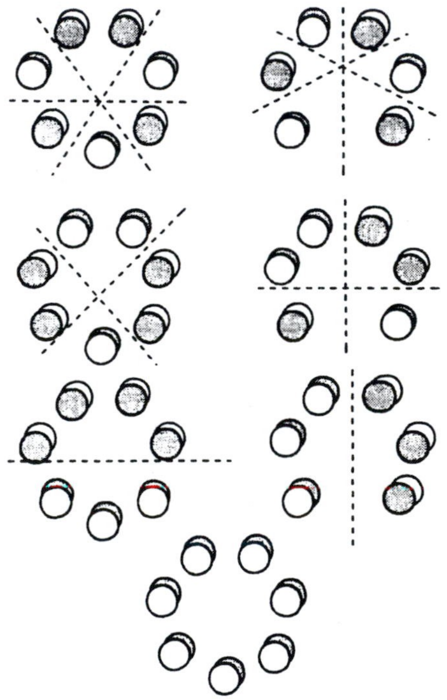

13-15 a. (η^4-C_4H_4)$Mo(CO)_4$ requires use of the group theoretical method described in Chapter 4. The four CO ligands are in a square planar arrangement, but the Mo atom is above them. The reducible representation $\Gamma = A_1 + B_1 + E$, with A_1 and E IR active, so there are two bands visible (the E bands are degenerate).

C_{4v}	E	$2C_4$	C_2	$2\sigma_v$	$2\sigma_d$	
Γ	4	0	0	2	0	
A_1	1	1	1	1	1	z
B_1	1	-1	1	1	-1	
E	2	0	-2	0	0	(x, y)

b.

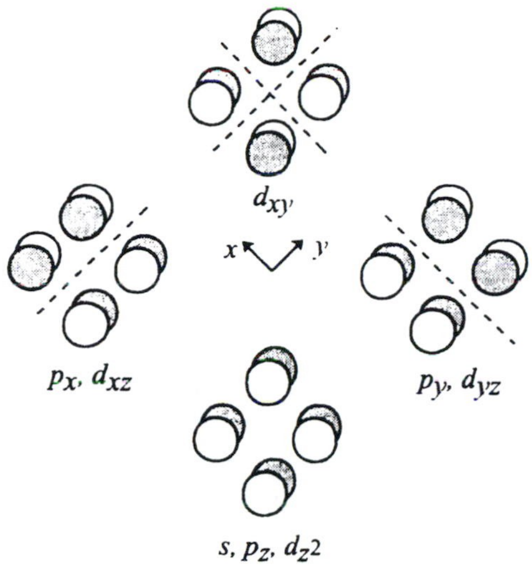

13-16 a. From bottom to top:

0-node, left, is A_1', 0 node, right, is A_2''

1-node, far left and third from left are E_1'

1-node, second from left and far right are E_1''

2-node, far left and third from left are E_2'

2-node, second from left and far right are E_2''

b.

s, d_{z^2}	$(d_{xy}, d_{x^2-y^2})$	(d_{xz}, d_{yz})	p_z	(p_x, p_y)
A_1'	E_2'	E_1''	A_2''	E_1'

c. The matching orbitals are shown in b; there is no match in the Fe orbitals for the E_2'' ligand orbitals.

13-17 a. The π orbitals of benzene are shown in Figure 13-22, page 482.

b. Group orbitals

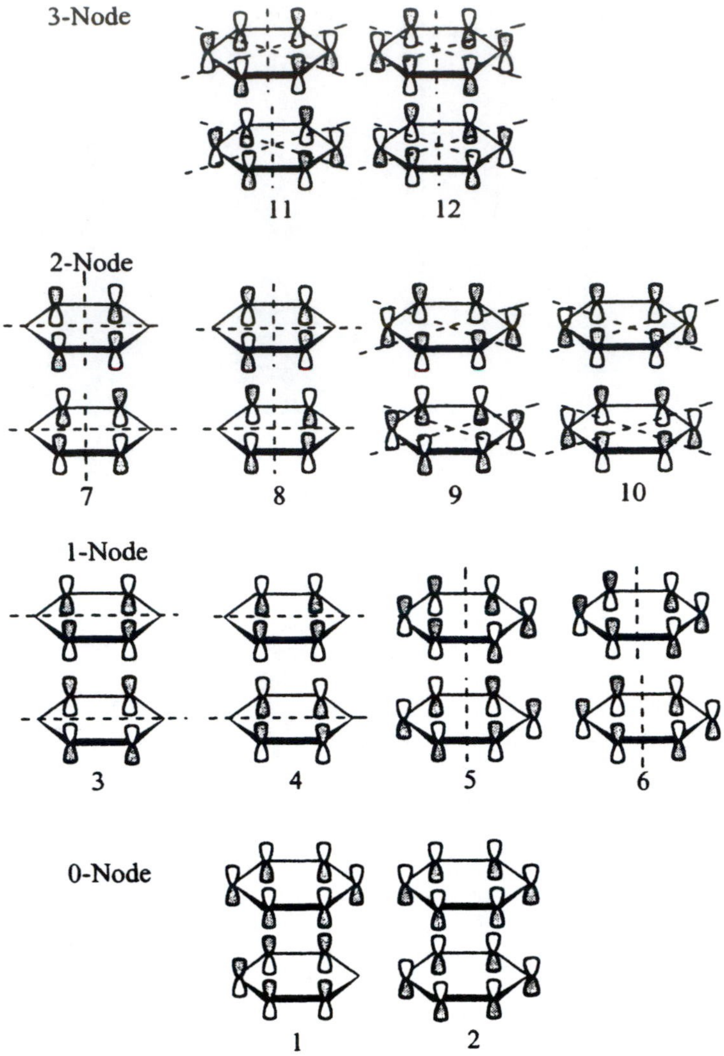

c. Matching Cr orbitals:

1	s, d_{z^2}	7	d_{xy}
2	p_z	8	none
3	p_y	9	$d_{x^2-y^2}$
4	d_{yz}	10	none
5	p_x	11	none
6	d_{xz}	12	none

d. Energy level diagram

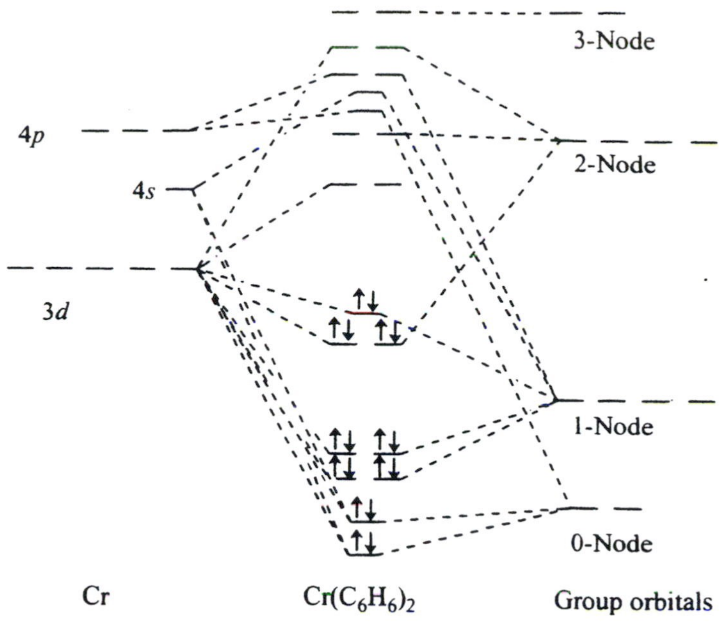

13-18

C_{3v}	E	$2C_3$	$3\sigma_v$	
Γ	3	0	1	
A_1	1	1	1	z
A_2	1	1	-1	
E	2	-1	0	(x, y)

$\Gamma = A_1 + E$, all IR active. Three vibrations, but two are degenerate, so two C-O stretching bands are expected in the IR spectrum.

13-19. $W(CO)_5(NCC_3H_7)$ 2077, 1975, 1938 cm^{-1}

$W(CO)_4(NCC_3H_7)_2$, *cis* 2107, 1898, 1842 cm^{-1}

$W(CO)_3(NCC_3H_7)_3$, probably *fac* 1910, 1792 cm^{-1}

a. Pentacarbonyls have three active IR stretches, tetracarbonyls have 1 (other ligands *trans*) or 4 (other ligands *cis*), and tricarbonyls have 2 (*fac*) or 3 (*mer*). It appears that two of the tetracarbonyl bands have the same energy, so only 3 show, but that the nitriles must be *cis*. The evidence for the *fac* isomer of the tricarbonyl is not conclusive, since again two bands of the *mer* isomer might have identical energies.

b. The stretching energy of the CO *trans* to the nitrile ligand is lower than that for CO *cis* due to reduced competition for the π back-bonding electrons (CO is a better π-acceptor). In the second complex, two are *cis* and two are *trans*. In the third complex, all three are *trans*, so the band energy is lower. In general, the trend is to lower energies as CO is replaced by butyronitrile; butyronitrile is not as effective a π acceptor as CO, so the π-acceptor nature of CO is enhanced as more nitrile ligands are added.

13-20 a. The bands at lower energy are for the CO ligands. A *cis* dicarbonyl octahedral complex should have two C—O stretches in the infrared spectrum (both symmetric and antisymmetric stretches are IR active), a *trans* dicarbonyl complex should have a single C—O stretch in the IR (the symmetric stretch is not IR active), and a monocarbonyl complex should of course have a single C—O stretch in the IR. These behaviors match the bands on the right (lower energy) in the spectra shown. Also, CO is a much better π acceptor ligand than CN^-, so it is not surprising that the C—O stretches occur at lower energies than C—N stretches in these ions. Finally, CO has a slightly greater reduced mass than CN^- and therefore has lower energy vibrational modes (the energy of vibrational levels is inversely proportional to $\sqrt{\text{reduced mass}}$).

b.

Complex	C—O stretches	C—N stretches
2: *cis*-$[Fe(CO)_2(CN)_4]^{2-}$	2	4
1: *trans*-$[Fe(CO)_2(CN)_4]^{2-}$	1	1
3: *cis*-$[Fe(CO)(CN)_5]^{3-}$	1	3

In compounds 2 and 3 some of the C—N stretches are too weak to be seen or the bands overlap; otherwise, the numbers of expected bands and the actual spectra match.

13-21 a. The CO ligand can be in either the axial or the equatorial position on the trigonal bipyramidal complex. The electronic environments are not identical, so the two isomers should have different carbonyl stretching bands.

b. The CO stretch absorbs at a higher energy in $Fe(CO)(PF_3)_4$, which means that the C-O bond is stronger in that compound. If the CO bond is stronger, the Fe-C bond is correspondingly weaker, which means that the PF_3 ligands are better π acceptors and therefore higher in the spectrochemical series than CO.

13-22

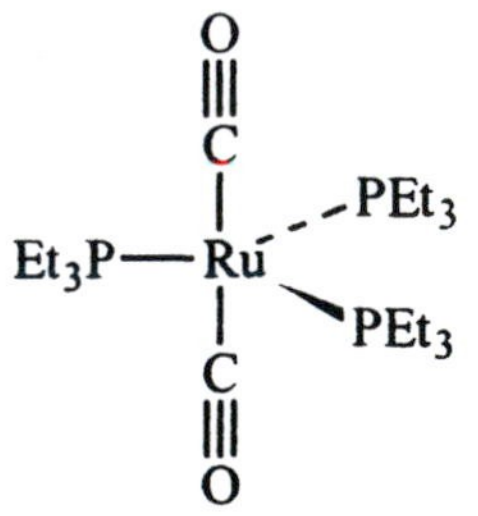

One C-O stretch (symmetric) observable at $\nu_{CO} = 1867\ cm^{-1}$

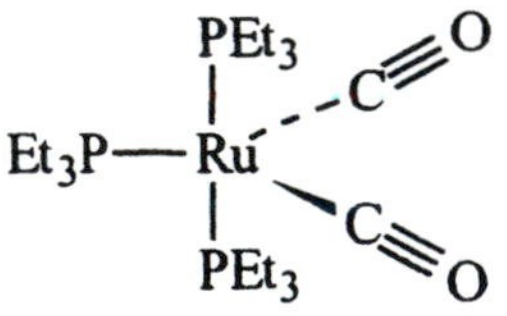

Two C-O stretches (symmetric and antisymmetric) should be observable. Only one seen at $\nu_{CO} = 1850\ cm^{-1}$

Calculations reported by the authors indicate that the second band of the *cis* isomer is likely to be very weak.

13-23 From Table 13-7 (p. 505), $[Co(CO)_3(PPh_3)_2]^+$ must be trigonal bipyramidal with all three CO ligands in the equatorial plane, which results in a single C-O absorption band.

13-24 By extrapolation of the positions of the C—O bands for $[Mn(CO)_6]^+$ (2100 cm^{-1}) and $[Fe(CO)_6]^{2+}$ (2204 cm^{-1}), one might predict a comparable band for $[Ir(CO)_6]^{3+}$ near 2300 cm^{-1}. Actual value: 2254 cm^{-1}.

13-25 a. In all three cations the strongly positive metal pulls electrons strongly from the CO ligands, significantly reducing backbonding. This effect is greatest for $[Hg(CO)_2]^{2+}$, with a charge of 1+ per CO, and weakest for $[Os(CO)_6]^{2+}$, with a charge of 1/3+ per CO. Therefore, $[Os(CO)_6]^{2+}$ should have the strongest backbonding, weakest C—O bond, and lowest energy carbon—oxgyen stretching vibration. Actual positions of IR bands:

	ν(CO), cm^{-1}
$[Hg(CO)_2]^{2+}$	2278
$[Pt(CO)_4]^{2+}$	2244
$[Os(CO)_6]^{2+}$	2190

b. A reducible representation for this ion based on C—O stretching vibrations would have the following characters:

	E	$2S_4$	C_2	$2C_2'$	$2\sigma_d$	
Γ	6	0	2	0	4	
$2\,A_1$	2	2	2	2	2	
$2\,B_2$	2	-2	2	-2	2	z
E	2	0	-2	0	0	(x, y)

There should be three carbon-oxygen stretching bands, two of B_2 symmetry and one of E symmetry (a degenerate pair). With a charge of 2+, this complex should exhibit infrared bands near those of $[Fe(CO)_6]^{2+}$ (2204 cm^{-1}). Actual bands are observed at 2173 cm^{-1} (E) and at 2187 and 2218 cm−1 (B_2).

13-26 $(\eta^5\text{-}C_5H_5)Cr(CO)_2(NS)$ 1962, 2033 cm^{-1}

$(\eta^5\text{-}C_5H_5)Cr(CO)_2(NO)$ 1955, 2028 cm^{-1}

NS is a stronger π-acceptor, so the CO back-bonding is reduced in $(\eta^5\text{-}C_5H_5)Cr(CO)_2(NS)$ and the C≡O bond is strengthened and has a higher stretching energy.

13-27 Because the reduced mass of ^{13}CO is greater than the reduced mass of ^{12}CO, the separation between vibrational energy levels should be less for ^{13}CO, and the ^{13}CO complex should therefore show an infrared band at lower energy than 2199 cm^{-1}. Actual value: 2149 cm^{-1}

13-28 a.

$$Mo(CO)_6 + Ph_2PCH_2PPh_2 \xrightarrow{\Delta} \underset{\underset{Ph_2P}{|}}{\overset{\overset{Ph_2P}{|}}{H_2CMo(CO)_4}} + 2\,CO$$

b. $(\eta^5\text{-}C_5H_5)(\eta^1\text{-}C_3H_5)Fe(CO)_2 \xrightarrow{h\nu} (\eta^5\text{-}C_5H_5)(\eta^3\text{-}C_3H_5)Fe(CO) + CO$

The allyl ligand can bond in either η^1 or η^3 fashion. Loss of CO converts the compound from 18 electrons to 16 electrons; rearrangement of allyl from η^1 to η^3 returns it to 18 electrons.

c. $(\eta^5\text{-}C_5Me_5)Rh(CO)_2 \longrightarrow [(\eta^5\text{-}C_5Me_5)Rh(CO)]_2 + 2CO$
A double Rh–Rh bond is needed; COs may be bridging or not. The electron count is the same either way.

d. $V(CO)_6 + NO \longrightarrow V(CO)_5(NO) + CO$ The compound changes from 17 to 18 electrons.

e. $W(CO)_5=C(C_6H_5)(OC_2H_5) + BF_3 \longrightarrow [(CO)_5W\equiv CC_6H_5]^+ + F^- + F_2BOC_2H_5$

f. $[(\eta^5\text{-}C_5H_5)Fe(CO)_2]_2 + 2\,Al(C_2H_5)_3 \longrightarrow$

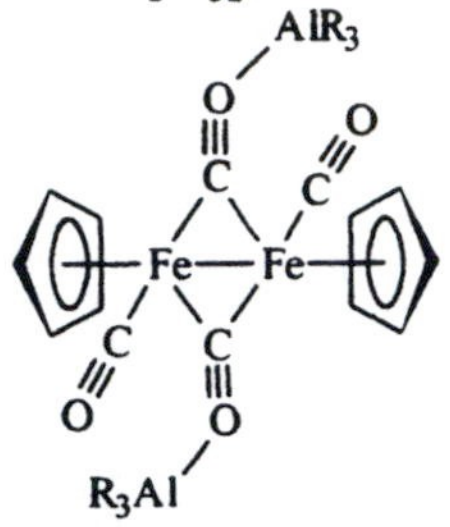

13-29

	$\nu(CO)$, cm^{-1}
$P(t\text{-}C_4H_9)_3$	1923 (best σ donor, poorest π acceptor)
$P(p\text{-}C_6H_4Me)_3$	1965
$P(p\text{-}C_6H_4F)_3$	1984
$P(C_6F_5)_3$	2004 (best π acceptor)

13-30 a. $Fe(CO)_4(PF_3)$ (PF_3 is a strong π acceptor)

b. $[Re(CO)_6]^+$

c. $Mo(CO)_3(PCl_3)_3$ (PCl_3 is the best π acceptor among the phosphines)

13-31   In order from highest to lowest ν:
$Mo(CO)_4(F_2PCH_2CH_2PF_2)$
$Mo(CO)_4((C_2F_5)_2PCH_2CH_2P(C_2F_5)_2)$
$Mo(CO)_4((C_6F_5)_2PCH_2CH_2P(C_6F_5)_2)$
$Mo(CO)_4(Ph_2PCH_2CH_2PPh_2)$
$Mo(CO)_4(Et_2PCH_2CH_2PEt_2)$

13-32   Coordinated N_2 has a lower stretching energy than free N_2. N_2 can act as a π acceptor, weakening the N–N bond and lowering the energy of the stretching vibration. The stretching vibration of free N_2 is not seen in the IR spectrum because there is no change in dipole moment on stretching.

13-33 At the higher temperature, the $-O-CH_3$ group should rotate rapidly enough to show only a single, average electronic environment, so each CH_3 has a single peak. At low temperatures, this rotation can be restricted, and *cis* and *trans* isomers result (see Figure 13-42). The larger peaks represent the more prevalent isomer.

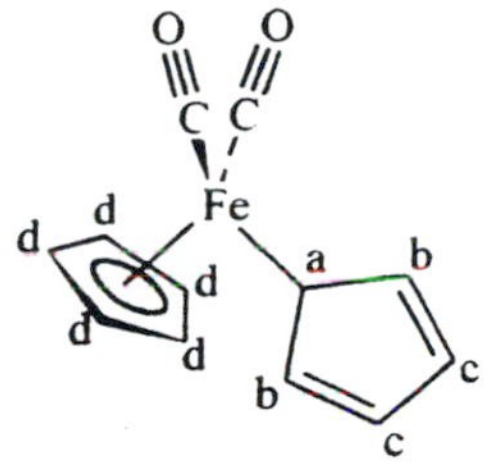

13-34 At high temperature the C_5H_5 rings undergo rapid 1,2 shifts to give a single average proton signal on the NMR time scale. At sufficiently low temperature, the different environments of the η^1 and η^5 rings can be seen. The relative intensities are a = 1, b = 2, c = 2, and d = 5.

13-35 PF_3 is a better π acceptor than PCl_3. As a result, the chromium in $Cr(CO)_5(PCl_3)$ has a greater electron density and CO acts as a stronger π acceptor in this complex. As a result:

a. $Cr(CO)_5(PF_3)$ has the stronger and shorter C–O bonds.

b. $Cr(CO)_5(PCl_3)$ has the higher energy Cr–C bands; since CO acts as a better π acceptor in this complex, the Cr–C bond is strengthened.

13-36 a. $[Fe(NO)(mnt)_2]^-$ has less electron density on Fe, less backbonding to NO, and a stronger N–O bond with higher stretching frequency.

b. $(CO)_5Cr{:}N{\equiv}N{:}Cr(CO)_5$ N_2 acts as a π acceptor toward both metals, significantly weakening the N–N bond.

c. $Ta{=}CH_2$ has a shorter Ta–C bond because it is a double bond.

d. $Cr{\equiv}CCH_3$ has a triple bond, which is shorter than either of the Cr–C carbonyl bonds.

e. $[Fe(CO)_4]^{2-}$ has more back π bonding because Fe has the lowest nuclear charge of the metals in this isoelectronic series; this reduces the C≡O bonding and the energy of the C–O vibration.

13-37 $Ni(CO)_4$

T_d	E	$8C_3$	$3C_2$	$6S_4$	$6\sigma_d$	
Γ	4	1	0	0	2	
A_1	1	1	1	1	1	
T_2	3	0	−1	−1	1	(x, y, z)

$\Gamma = A_1 + T_2$ A_1 is IR inactive, T_2 is IR active: 1 band.

$Cr(CO)_6$

O_h	E	$8C_3$	$6C_2$	$6C_4$	$3C_2$	i	$6S_4$	$8S_6$	$3\sigma_h$	$6\sigma_d$	
Γ	6	0	0	2	2	0	0	0	4	2	
A_{1g}	1	1	1	1	1	1	1	1	1	1	
E_g	2	-1	0	0	2	2	0	-1	2	0	
T_{1u}	3	0	-1	1	-1	-3	-1	0	1	1	(x, y, z)

$\Gamma = A_{1g} + E_g + T_{1u}$ Only T_{1u} is IR active: 1 band

13-38

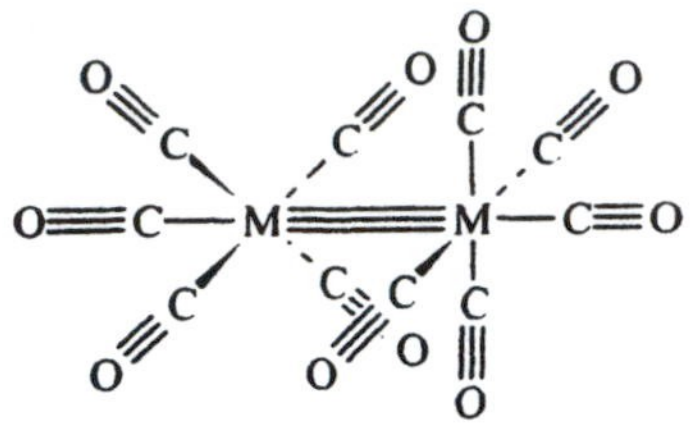

D_{4d}	E	$2S_8$	$2C_4$	$2S_8^3$	C_2	$4C_2'$	$4\sigma_d$	
Γ	10	0	2	0	2	0	4	
A_1	1	1	1	1	1	1	1	
B_2	1	-1	1	-1	1	-1	1	z
E_1	2	$\sqrt{2}$	0	$-\sqrt{2}$	-2	0	0	(x, y)
E_2	2	0	-2	0	2	0	0	
E_3	2	$-\sqrt{2}$	0	$\sqrt{2}$	-2	0	0	(R_x, R_y)

$\Gamma = 2A_1 + 2B_2 + E_1 + E_2 + E_3$ $2B_2$ and E_1 are IR active: 3 bands

13-39 One CO is replaced by 2-butyne. The NMR peaks are due to ethyl CH_3 ($\delta = 0.90$), ethyl CH_2 ($\delta = 1.63$), and butyne CH_3 ($\delta = 3.16$). The $\delta = 3.16$ peak splits at low temperatures because the two ends of the butyne are not identical; at higher temperatures, they become identical on the NMR time scale, perhaps through rotation about the Mo-butyne bond. The single ^{31}P peak indicates identical PEt_3 groups, suggesting the isomer shown. The IR indicates a carbonyl remains on the compound; the molecular weight of the compound shown is 574.2, well within the limits given.

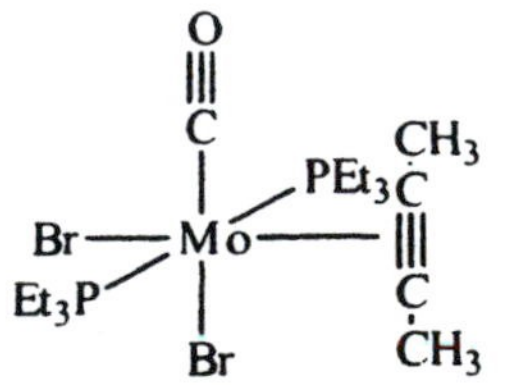

13-40 The analysis fits $[(\eta^5\text{-}C_5H_5)Fe(\mu\text{-}CO)_3Fe(\eta^5\text{-}C_5H_5)]$. The single CO band means that the three carbonyls must be identical:

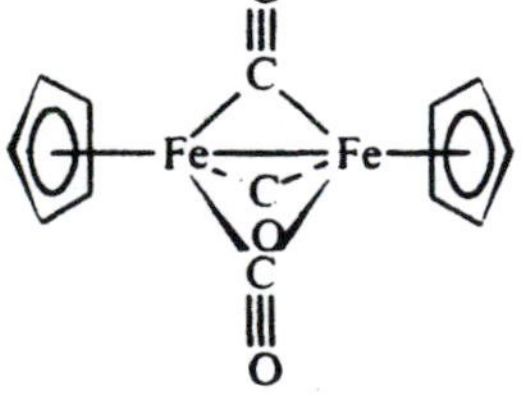

13-41 $(\eta^5\text{-}C_5H_5)(\eta^3\text{-}C_5H_7)Ni$ There are five η^5 protons, four on the two carbons of the second Cp ring that are not bonded to Ni, two on the first and third carbons bonded to Ni, and one on the center C bonded to Ni.

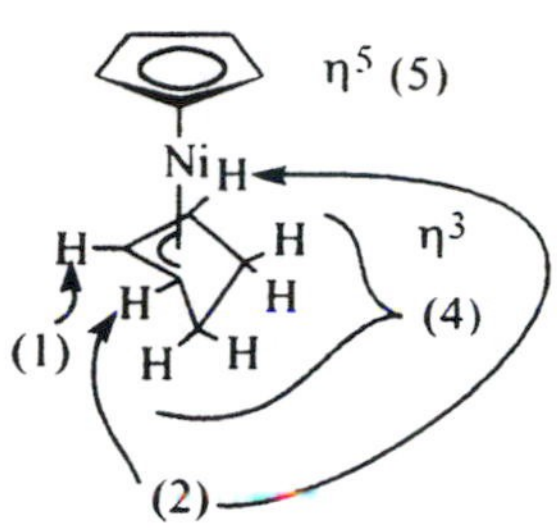

13-42 $I_s/I_a = \cot^2(\phi/2) = \cot^2 38° = 1.64$

13-43 **a.**

Mass of C_{60}:	720	
Ir:	193	(most abundant isotope)
CO:	28	
C_9H_7:	115	
	1056	

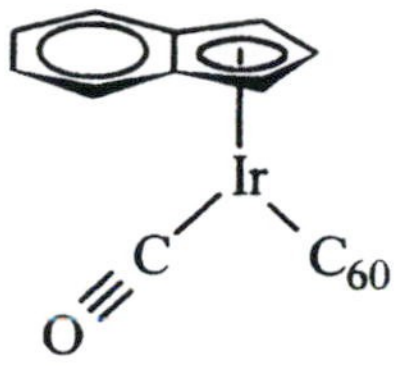

b. Since the carbonyl stretch decreases by 44 cm^{-1}, there must be a significant decrease in the electron density at Ir.

c. The C_{60} is replaced by PPh_3, giving $(\eta^5\text{-}C_9H_7)Ir(CO)(PPh_3)$.

13-44 If CO is liberated from $(\eta^5\text{-}C_5H_5)Mn(CO)_3$ it must be replaced by some other ligand if the 18-electron rule is to be maintained. In this case, the new ligand is tetrahydrofuran (THF), a cyclic ether that can act as a sigma donor, and compound **Q** is $(\eta^5\text{-}C_5H_5)Mn(CO)_2(THF)$:

$$(\eta^5\text{-}C_5H_5)Mn(CO)_3 + THF \rightarrow \underset{\textbf{Q}}{(\eta^5\text{-}C_5H_5)Mn(CO)_2(THF)} + CO$$

The NC groups on carbon in $H_2C(NC)_2$ can act as donors to transition metals; each has a lone pair on carbon. In the formation of **R** the weakly bound THF is replaced by a $H_2C(NC)_2$ ligand:

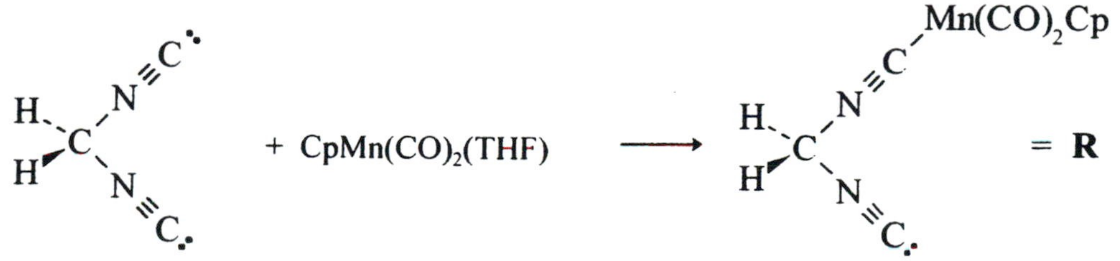

13-45 There are 0.00300 mmol of dimers in the solution. Of the metals in these dimers, one third are Mo and two thirds are W.

The probability that a particular molecule has formula $[CpMo(CO)_3]_2$ is = 1/3 × 1/3 = 1/9. Therefore, 1/9 × total moles = 1/9 × 0.00300 mmol = 0.00033 mmol $[CpMo(CO)_3]_2$.

The probability that a molecule has formula $[CpW(CO)_3]_2$ =
2/3 ×2/3 = 4/9, and 4/9 × 0.00300 mmol = 0.00133 mmol $[CpW(CO)_3]_2$.

The probability that a molecule has formula $Cp(CO)_3Mo–W(CO)_3Cp$ = 2 × 1/3 × 2/3 = 4/9, and 4/9 × 0.00300 mmol = 0.00133 mmol $Cp(CO)_3Mo–W(CO)_3Cp$.

13-46 The IR bands at 1945 and 1811 cm^{-1} are similar to those of $[Ti(CO)_4(\eta^5\text{-}C_5H_5)]^-$, suggesting similarities in structure and charge between this Ti complex and **Z**. If the other IR bands are for B—H stretches and the ^1H NMR shows two signals of relative area 3:1, it is reasonable to suggest that the boron might be present in the BH_4^- ion, with three of the hydrogens in one environment, the fourth in another. The peak at 2495 cm^{-1} is distinctly different in energy from the peaks at 2132 and 2058 cm^{-1} and might correspond to a stretch involving the less abundant of the protons. All is consistent with a structure having 4 CO ligands and one BH_4^- ligand (which has replaced the cyclopentadienyl ligand), and a negative charge.

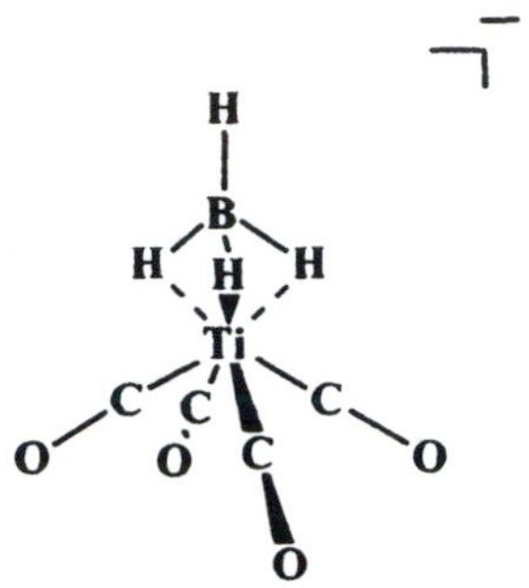

Symmetry analysis indicates that this complex should have two C—O stretching bands in the infrared (A_1 and E in local C_{4v} symmetry) and three B—H bands (two A_1 and one E in local C_{3v} symmetry). The higher energy B—H stretch is for the terminal H; the lower energy bands are for the bridging H atoms.

13-47 Different software and different parameter settings will generate orbitals with slightly different shapes and energies than those shown in the text, although the results should be similar. The relative energies in part **c** in particular may differ from those in the text, because this calculation is sensitive to the methods used. In part **d** the d_{z^2} conical nodal surface of the orbital is close to the p orbitals of the C_5H_5 rings, so the interaction between this orbital and the rings is weak.

13-48 Donor interaction (from π orbital of ethylene):

Acceptor interaction (involving π* of ethylene):

13-49 The shapes of the orbitals should be similar to those in Figure 13-8. The e_g* orbitals have different shapes because the d orbitals involved (d_{z^2} and $d_{x^2-y^2}$) have different shapes.

Chapter 14
Organometallic Reactions and Catalysis

14-1 **a.** $[Mn(CO)_5]^- + H_2C{=}C{-}CH_2{-}Cl \rightarrow H_2C{=}C{-}CH_2{-}Mn(CO)_4 + Cl^-$
$H_2C{=}C{-}CH_2{-}Mn(CO)_5 \rightarrow CO + (\eta^3\text{-}C_3H_5)Mn(CO)_4$

b. Oxidative addition

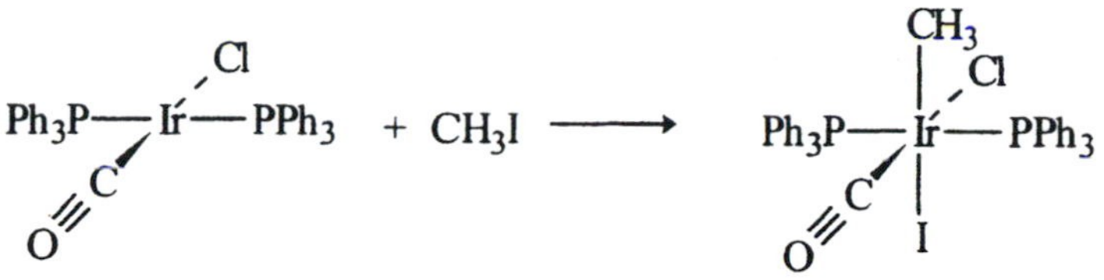

c. Cyclometallation

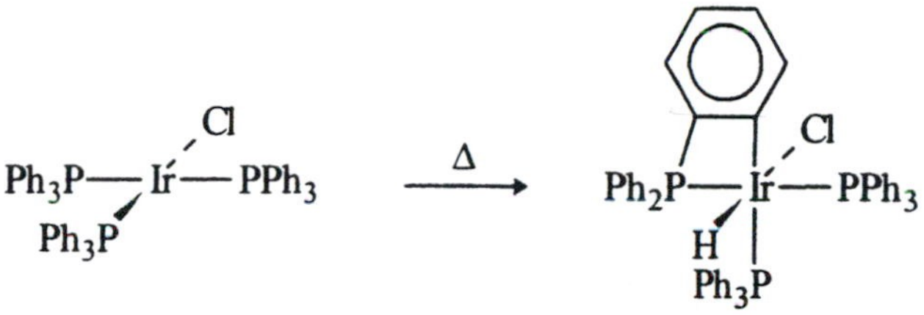

d. Two possibilities: methyl migration and addition or substitution

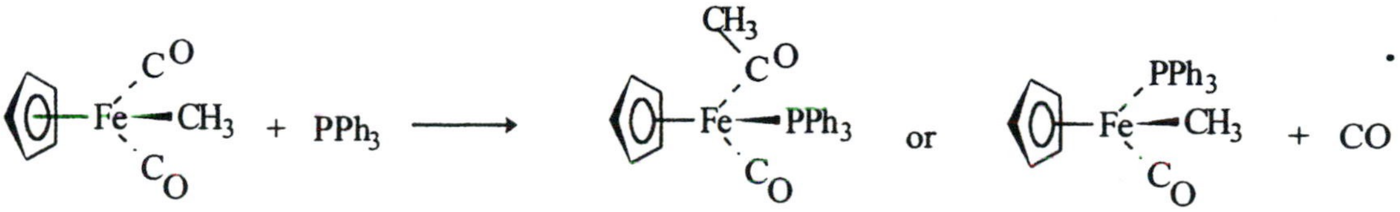

e. $(\eta^5\text{-}C_5H_5)Mn(CO)_3[C({=}O)CH_3] \longrightarrow (\eta^5\text{-}C_5H_5)Mn(CO)_3(CH_3) + CO$ (dissociation and methyl migration)

f. $H_3C\text{-}Mn(CO)_5 + SO_2 \longrightarrow H_3C\text{-}S({=}O)_2\text{-}Mn(CO)_5$ (1,1 insertion)

14-2 **a.** $H_3C\text{-}Mn(CO)_5 + P(CH_3)(C_6H_5)_2 \longrightarrow H_3C(C{=}O)\text{-}Mn(CO)_4[P(CH_3)(C_6H_5)_2]$
(methyl migration and phosphine addition)

b. $[Mn(CO)_5]^- + (\eta^5\text{-}C_5H_5)Fe(CO)_2Br \longrightarrow (CO)_5Mn{-}Fe(\eta^5\text{-}C_5H_5)(CO)_2 + Br^-$
(nucleophilic displacement)

c. *trans*-$Ir(CO)Cl(PPh_3)_2 + CH_3I \longrightarrow$

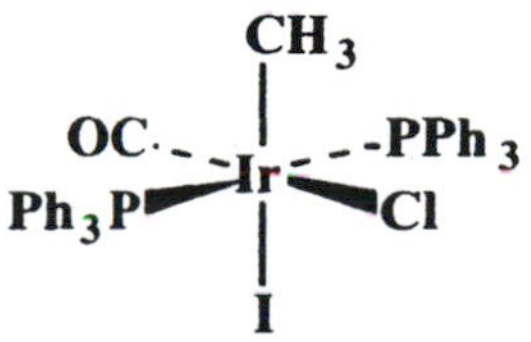

d. $W(CO)_6 + C_6H_5Li \longrightarrow [C_6H_5COW(CO)_5]^- + Li^+$ (nucleophilic attack on carbonyl C; see pp. 499-500)

e. Alkyl migration of CH_3 to one of the adjacent CO's, followed by addition of ^{13}CO. Isomeric products: 1/3 each of the *fac* enantiomers, 1/3 *mer* isomer. There should be no ^{13}C in the acyl group.

f. Alkyl migration of CH_3 to one of the adjacent CO's, followed by addition of ^{13}CO. Isomeric products: two *mer* species, enantiomers if the ^{13}CO is taken into account, identical otherwise.

14-3 **a.** *cis*-$(^{13}CO)(CH_3CO)Mn(CO)_4 \longrightarrow Mn(CH_3)(CO)_5$, 25% with no ^{13}CO, 25% with ^{13}CO *trans* to CH_3, 50% with ^{13}CO *cis* to CH_3.

b. $C_6H_5\text{-}CH_2\,Mn(CO)_5 \xrightarrow{h\nu} CO +$ 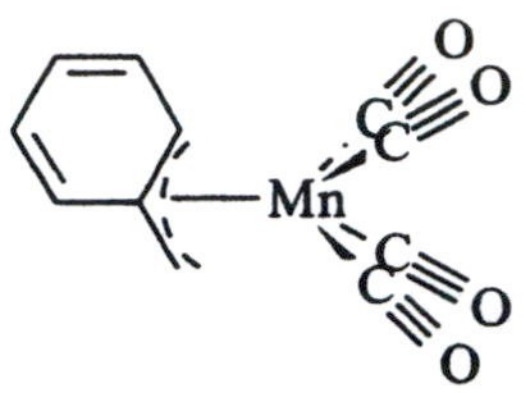

c. $[V(CO)_6] + NO \longrightarrow [V(CO)_5(NO)] + CO$

d. $Cr(CO)_6 + 2\ Na/NH_3 \longrightarrow 2\ Na^+ + [Cr(CO)_5]^{2-} + CO$

e. $Fe(CO)_5 + NaC_5H_5 \longrightarrow Na^+ + [(\eta^5\text{-}C_5H_5)Fe(CO)_2]^- + 3\ CO$

f. $[Fe(CO)_4]^{2-} + CH_3I \longrightarrow [(CH_3)Fe(CO)_4]^- + I^-$

g.

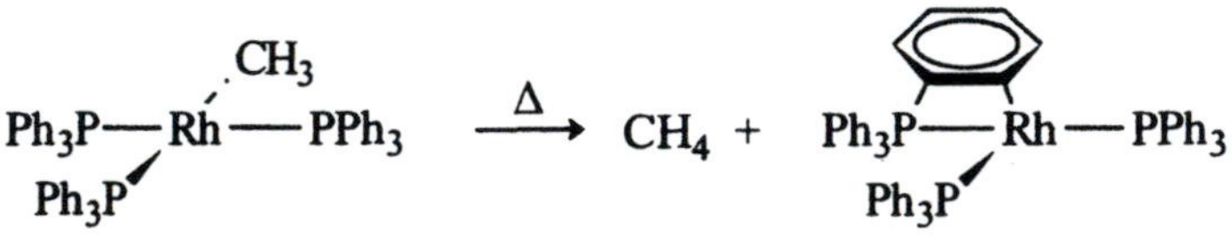

14-4 $[(C_5H_5)Fe(CO)_3]^+ + NaH \longrightarrow$ **A** $(C_7H_6O_2Fe)$ — **A** $= (C_5H_5)Fe(CO)_2H$

A → **B** (colorless gas) + **C** $(C_7H_5O_2Fe)$ (purple-brown) (see Figure 13-32, p. 458) — **B** $= H_2$; **C** $= [(C_5H_5)Fe(CO)_2]_2$

C $+ I_2 \longrightarrow$ **D** $(C_7H_5O_2FeI)$ (brown) — **D** $= (C_5H_5)Fe(CO)_2I$

D $+ TlC_5H_5 \longrightarrow$ **E** $(C_{12}H_{10}O_2Fe) + TlI$ — **E** $= (C_5H_5)_2Fe(CO)_2$, one Cp η^1 (Figure 13-32, p. 458)

E $\longrightarrow$ **F** $(C_{10}H_{10}Fe)$ + colorless gas (CO) — **F** $= (C_5H_5)_2Fe$, ferrocene

14-5 $[(\eta^5\text{-}C_5H_5)Fe(CO)_2]^- + ClCH_2CH_2SCH_3 \longrightarrow$ **A** $(C_{10}H_{12}FeO_2S)$

A + heat $\longrightarrow$ **B**

A has bands at 1980 and 1940 cm^{-1}, **B** at 1920 and 1630.

The sulfur reagent loses Cl^- and bonds to the Fe as an alkyl ligand to form **A**, with two carbonyls. This is an example of nucleophilic displacement of chloride by an organometallic anion. **A** then rearranges, with the S becoming attached to Fe, and the alkyl bonds to a carbonyl carbon. **B** contains an ordinary carbonyl and an acyl C=O bond, for the two quite different C-O stretching energies.

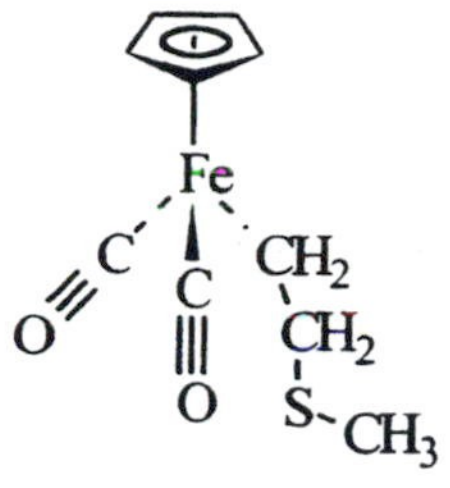

A

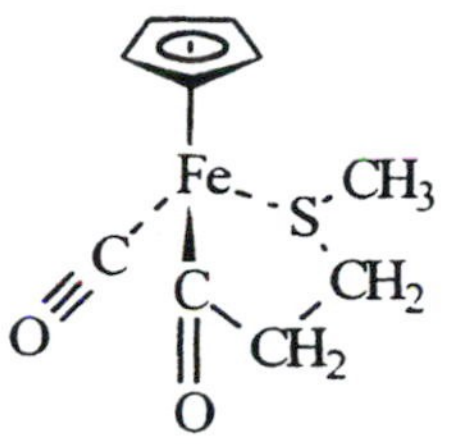

B

14-6 **a.** Two term rate laws like this could be the result of two parallel associative reactions, the first by solvent, the second by the phosphite, or could result from a dissociative reaction for the first term and an associative reaction for the second.

First term, dissociative, $k_1[V(CO)_5(NO)]$:

$$V(CO)_5(NO) \longrightarrow V(CO)_4(NO) + CO \xrightarrow{PR_3} V(CO)_4(NO)(PR_3)$$

Second term, associative, $k_2[PR_3][V(CO)_5(NO)]$:

$$V(CO)_5(NO) \xrightarrow{PR_3} V(CO)_5(NO)PR_3 \longrightarrow V(CO)_4(NO)(PR_3) + CO$$

b. $V(CO)_5[P(OCH_3)_3](NO)$ If the NO is a bent, 1-electron donor, it can be an 18-electron species $(5 + 5 \times 2 + 2 + 1 = 18)$. If NO is a linear, 3-electron donor, the total is 20 electrons.

14-7 $Co_2(CO)_8 \rightleftharpoons Co_2(CO)_7 + CO$ K_1 (fast equilibrium) $[Co_2(CO)_7] = K_1\ [Co_2(CO)_8]/[CO]$

$Co_2(CO)_7 + H_2 \longrightarrow Co_2(CO)_7H_2$ k_2 (slow)

$Co_2(CO)_7H_2 + CO \longrightarrow 2\ HCo(CO)_4$ (fast)

$$\text{Rate} = k_2\ [Co_2(CO)_7]\ [H_2] = k_2\ K_1\ [Co_2(CO)_8][H_2]/[CO]$$

14-8 This depends on the cone angle of the phosphine ligands, with the order $PPh_3 > PBu_3$ (estimated) $> P(OPh)_3 > P(OMe)_3$ from Table 14-1 (p. 523). The PPh_3 should dissociate most rapidly and the $P(OMe)_3$ should dissociate least rapidly.

14-9 *K* for the dissociation reaction is in the order $PPh_3 > PMePh_2 > PEt_3 > PMe_3$, as a result of a combination of decreasing cone angle and increasing negative charge on the phosphorus. Alkyls push more electron density onto P than phenyl rings.

14-10 If it loses CO followed by migration of CH_3 to an adjacent position, all the CO lost should be ^{12}CO, because the COs *cis* to the $CH_3C{=}O$ will be the ones lost.

14-11 **a.** There are two possible products, both a result of methyl migration followed by carbonyl addition. The methyl group can move to either the 1 position or the 2 position. The resulting products are different only in the location of the $^{13}C{\equiv}O$.

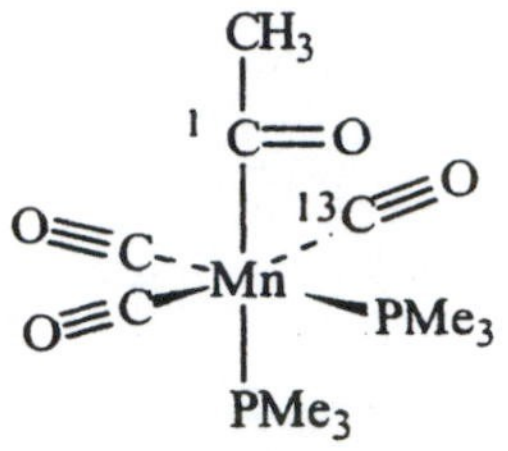

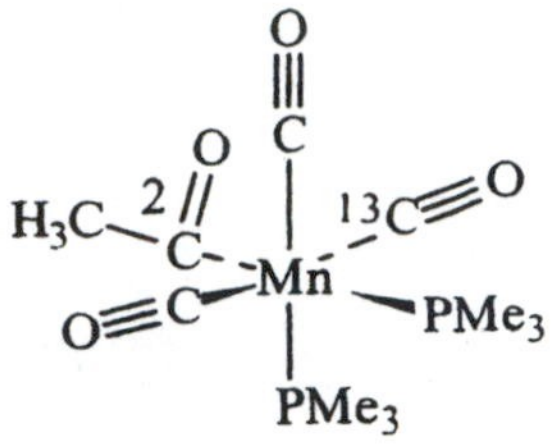

b. The new IR band is from the acyl carbonyl, and should be near 1630 cm^{-1} (see problem 14-5).

14-12 $[(C_5H_5)_2Fe_2(CO)_4]$ + Na/Hg ⟶ **A**
A + Br_2 ⟶ **B**
B + $LiAlH_4$ ⟶ **C**
C + PhNa ⟶ **A** + **D**
$[(C_5H_5)_2Fe_2(CO)_4]$ 1961, 1942, 1790 cm^{-1} 2 bridging CO, Fe-Fe single bond
A 1880, 1830 cm^{-1}
$Na^+[(C_5H_5)Fe(CO)_2]^-$
B $[(C_5H_5)Fe(CO)_2Br]$
C NMR δ = -12 ppm (1)
δ = 5 ppm (5) $[(C_5H_5)Fe(CO)_2H]$
D hydrocarbon (benzene, C_6H_6)

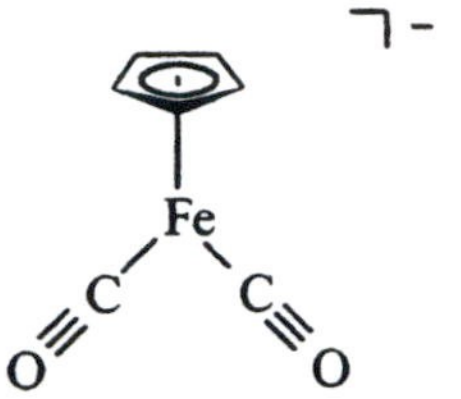

A

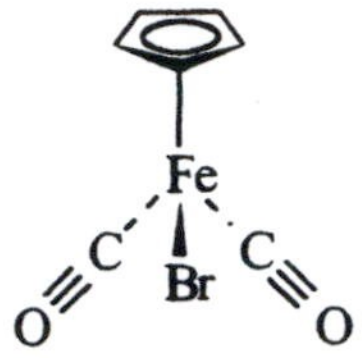

B

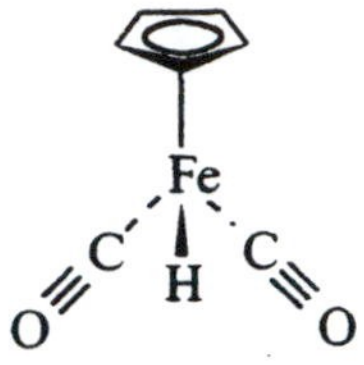

C

14-13 **a.** The bromoethoxide anion adds to a carbonyl. The C of the carbonyls are the hardest and most positive part of the complex. Then Br^- is lost, leaving a positive carbon. The alkyl tail can then bend around and react with the carbonyl oxygen, giving the compound shown in the figure.

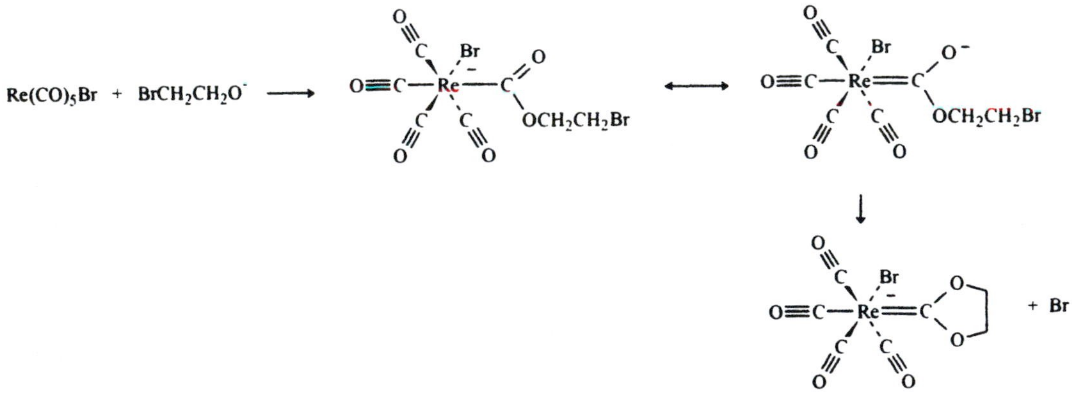

Y

b. There are two electrons donated from each of the six ligand positions, and Re^+ has 6 electrons for a total of 18. In the isomer shown, there are three different carbonyls, and the carbene ligand has two identical carbons, so there are five different magnetic environments for carbon. In the other possible isomer, with Br trans to the alkoxide ligand, there are not five different carbon environments. Finally, Ag^+ can remove Br^- from the complex.

14-14 a.   This is the example described on pages 511-512. See those pages for further details.

$I + PPh_3 \longrightarrow II$

$II + PPh_3 \longrightarrow III$

II   2038, 1958, 1906 cm^{-1}

          7.62, 7.41 multiplets (15), 4.19 multiplet (4)

III   1944, 1860 cm^{-1}

          7.70, 7.32 multiplets (15), 3.39 singlet (2)

II has 3 CO's in a *mer* geometry (see Table 13-7, p. 505), so the CO *trans* to Br has been replaced by PPh_3. The NMR shows one PPh_3 (15) and the ethylene hydrogens (4).

III has only 2 CO's in a *cis* configuration (Table 13-7), so an axial CO has been replaced. The ratio of NMR integrated peaks is now 15:2 because there are two PPh_3 groups. Why the C_2H_4 protons have collapsed into a single peak is unknown.

b. $I + Ph_2PCH_2PPh_2 \longrightarrow IV$

IV   2036, 1959, 1914 cm^{-1}, 35.87% C, 2.73% H

Replacing two CO's with the phosphine gives a compound with 25.7% C, 3.1% H. Replacing one CO and the Br with the phosphine gives a compound with 28.9% C, 3.3% H.

The only way to get the analysis to work out is to have a single diphos bridging two Re after loss of CO from each:

$[(Re(CO)_3Br(C(O_2C_2H_4))]_2\{(PPh_2)_2CH_2\}$ has 36.14% C, 2.46% H.

c. $I + S_2CN(CH_3)_2^- \longrightarrow Re(CO)_5Br + V$

V has no metal, has no IR bands between 1700 and 2300 cm^{-1}, has NMR bands at 3.91 (triplet), 3.60 (triplet), 3.57 (singlet), and 3.41 (singlet). This requires more NMR and IR analysis than is provided in this book, but regular organic sources show this to be $(CH_3)_2NC(=S)SC_2H_4O^-$, which will pick up a proton from any trace amounts of water to make the hydroxy compound. The 1500 and 977 cm^{-1} bands are the N-C and C-S bands, and the NMR shows triplets for the two CH_2 units, singlets for the CH_3's.

14-15   The mechanism is the one described in problem 13, with formation of the alkoxide ion by the reaction

$$\begin{array}{c} H_2C \\ | \\ H_2C \end{array}\!\!\!\!>\!\!O \ + \ Br^- \longrightarrow BrCH_2CH_2O^-$$

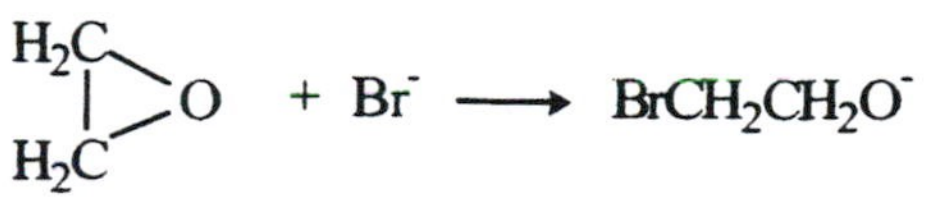

14-16   The reaction proceeds by substitution of $Mn(CO)_5^-$ for Br^-, followed by alkyl migration, addition of $Mn(CO)_5^-$, and finally cyclization:

$$Mn(CO)_5^- + BrCH_2CH_2CH_2Br \longrightarrow (CO)_5MnCH_2CH_2CH_2Br + Br^- \longrightarrow$$

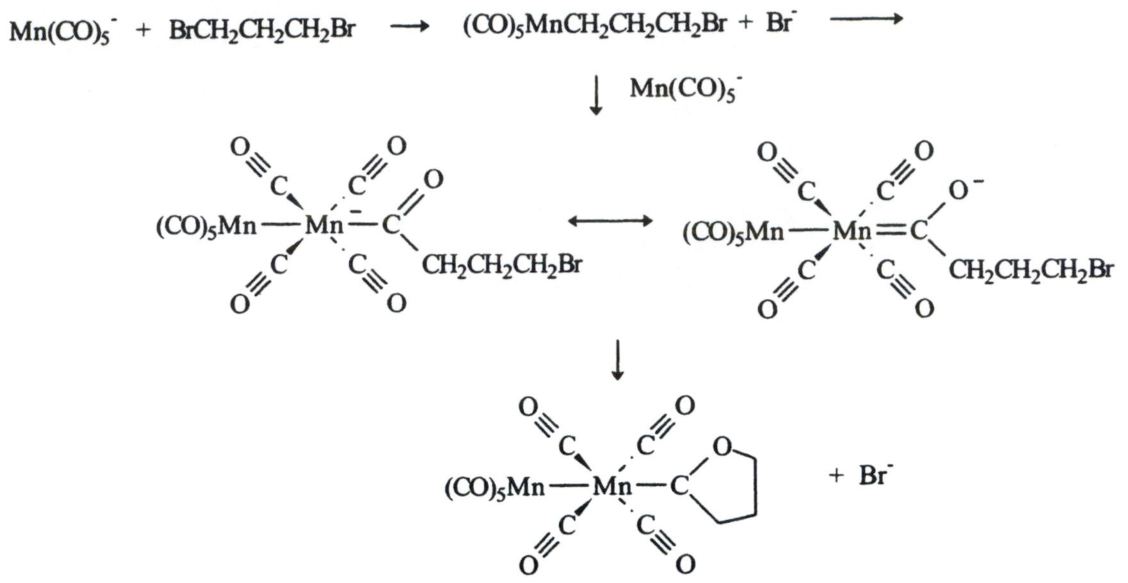

14-17   The acyl metal carbonyl has a resonance structure with a negative charge on oxygen. Protonation is then similar to the alkylation reaction described in Chapter 13 (p. 499).

14-18   **a.**   Acetaldehyde from ethylene:
The Wacker process (Figure 14-17, p. 541) with ethylene as the starting alkene will result in acetaldehyde.

$$C_2H_4 + CO + H_2 \xrightarrow{[PdCl_4]^{2-}} H_3C-\overset{\overset{\displaystyle O}{\|}}{C}H + H_2O$$

b.   Ethyl propionate from chloroethane:

$$Co(CO)_4^- \xrightarrow{CH_3CH_2Cl} CH_3CH_2Co(CO)_4 \xrightarrow{CO} (CH_3CH_2CO)Co(CO)_4 \xrightarrow{C_2H_5OH} CH_3CH_2COOC_2H_5 + HCo(CO)_4$$

c.   Pentanal from 1-butene: The hydroformylation process (Figure 14-14, p. 537) does this.

d.   4-phenylbutanal from an alkene: Again, the hydroformylation process should do this, starting with 4-phenyl-1-propene.

e.   Wilkinson's catalyst [ClRh(PPh₃)₃] (pp. 542) should do this. D_2 would be added across the least hindered double bond.

f. Catalytic deuteration with H_3TaCp_2 (Figure 14-13, page 536) should deuterate the phenyl ring without affecting the methyl hydrogens.

14-19 Figure 14-15, page 539 shows this process of hydroformylation. *n*-pentanal results if $R = C_2H_5$. Identification of the steps:

1. dissociation of one CO
2. addition of the alkene
3. 1,2 insertion
4. addition of CO
5. alkyl migration
6. oxidative addition
7. reductive elimination of the aldehyde

14-20 Hydroformylation (Figure 14-15, page 539) with $Rh(CO)_2(PPh_3)_2$ as the catalytic species will work, starting with 2-methyl-1-butene: $H_3C\text{-}CH_2\text{-}CH(CH_3)\text{=}CH_2$.

14-21 a. Direct metathesis would occur as follows:

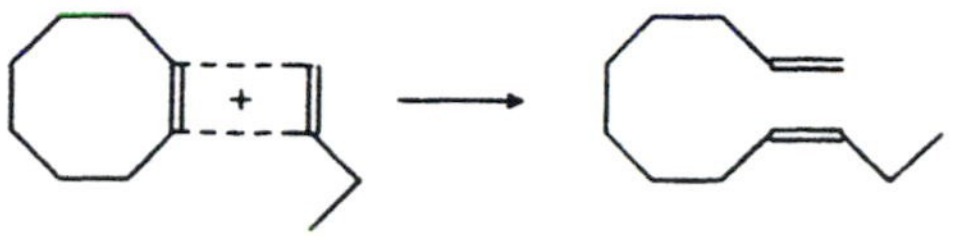

In addition, reactants could undergo self-metathesis. For example

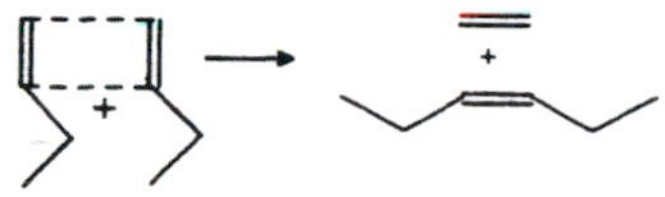

and the products of the direct metathesis could undergo further metathesis to give a variety of products. One of these could be formed as shown here.

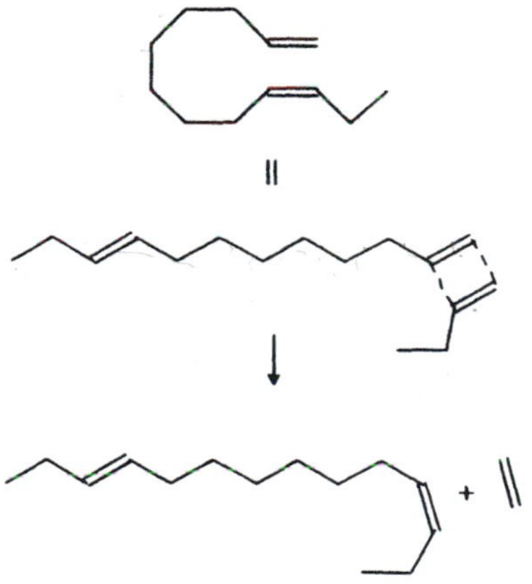

b.

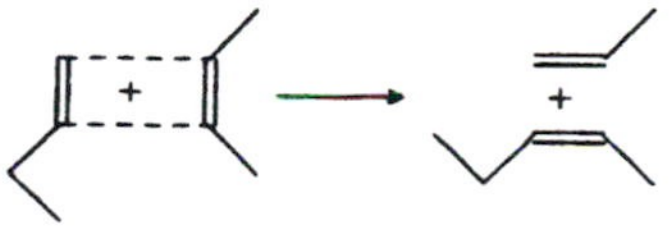

c.

14-22 **a.**

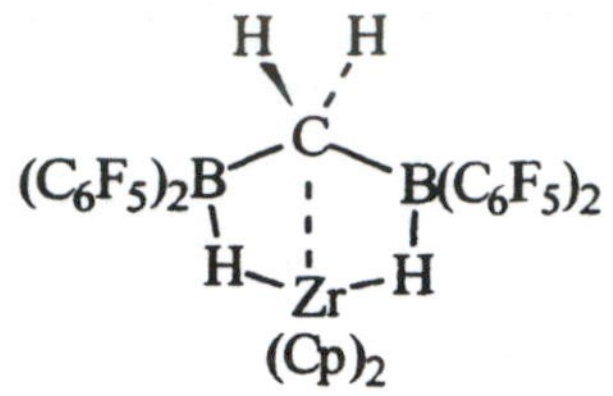

b. One possible route:

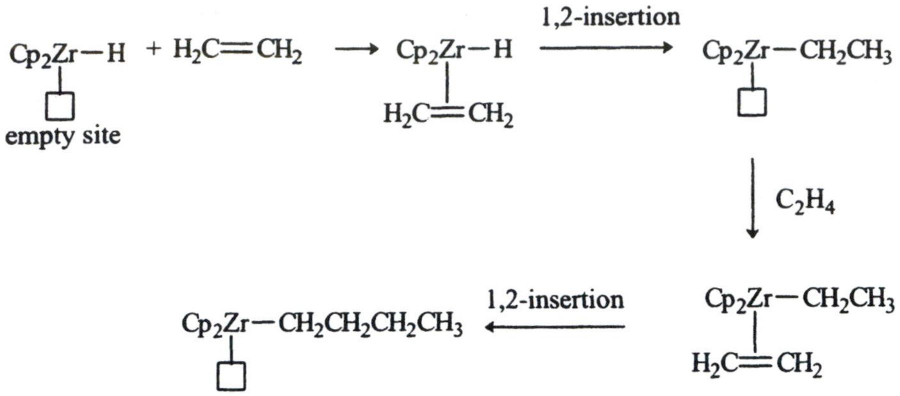

etc.

14-23

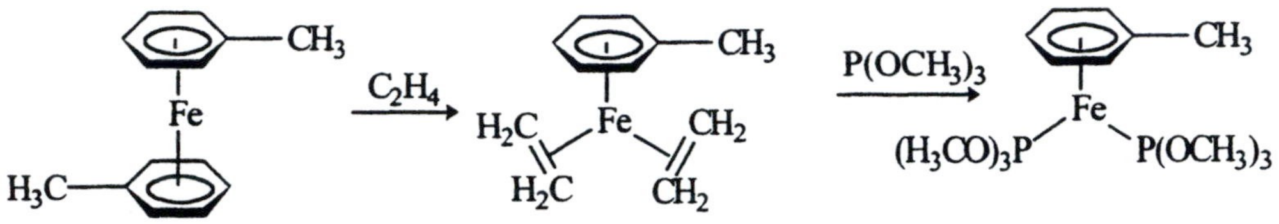

X

14-24 $RhCl_3 \cdot 3H_2O + P(o\text{-}MePh)_3 \rightarrow$ **I** (blue-green) ($C_{42}H_{42}P_2Cl_2Rh$) $\mu_{eff} = 2.3$ B.M., Rh-Cl $\nu = 351\ cm^{-1}$

I is *trans*-$RhCl_2(PR_3)_2$ The IR is the Rh-Cl asymmetric stretch. The *cis* isomer would give two IR bands. One unpaired electron (15 total, square planar).

I + heat $\rightarrow$ **II** (yellow, diamagnetic) Rh:Cl = 1, $\nu = 920\ cm^{-1}$

II Loss of one Cl and combination of two CH_3's to form a single tridentate phosphine ligand with a π bond. Sixteen electrons around Rh. The double bond in the ligand is perpendicular to the Rh-Cl-P plane because of the size and geometry of the benzene rings.

II + $SCN^- \rightarrow$ **III** Rh(SCN)

NMR: δ	Area	Type
6.9-7.5	12	aromatic
3.50	1	doublet of 1:2:1 triplets
2.84	3	singlet
2.40	3	singlet

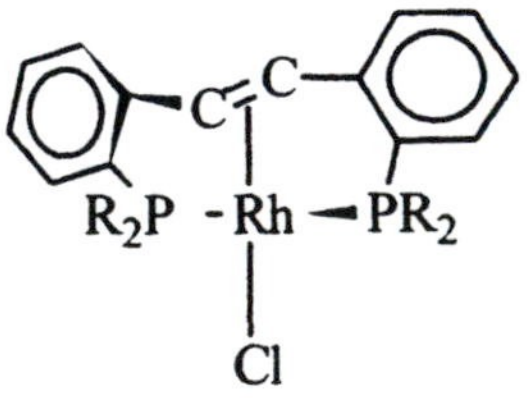

II

III Same as **II**, with SCN substituted for Cl. The singlets are the methyl protons (two separate environments--not obvious from the drawing), the doublet is from the vinylic protons, and the aromatic multiplets are the sum of all the phenyl protons.

II + NaCN → **IV** ($C_{21}H_{19}P$, mw = 604) ν = 965 cm^{-1}

NMR: δ	Area	Type
7.64	1	singlet
6.9-7.5	12	aromatic
2.37	6	singlet

IV is the diphosphine ligand shown above (CN^- substitutes for all the ligands in this reaction), actually $C_{42}H_{38}P_2$. The IR band is characteristic of *trans* vinylic hydrogens. All methyl protons are equivalent in the free ligand (singlet), the vinylic protons are a singlet at 7.64, and the phenyl protons are a multiplet at 6.9-7.5. The change in the vinyl hydrogen IR band shows that the ethylene does coordinate to Rh. Coordination reduces the C=C bonding, which also reduces the C–H bending energy (electrons are drawn away from C–H, toward C–Rh).

14-25 a. $[MeC{\equiv}C(CH_2)_2OOC(CH_2)_2]_2 = CH_3C{\equiv}CCH_2CH_2OOCCH_2CH_2COOCH_2CH_2C{\equiv}CCH_3$ (with C=O on the ester carbon)

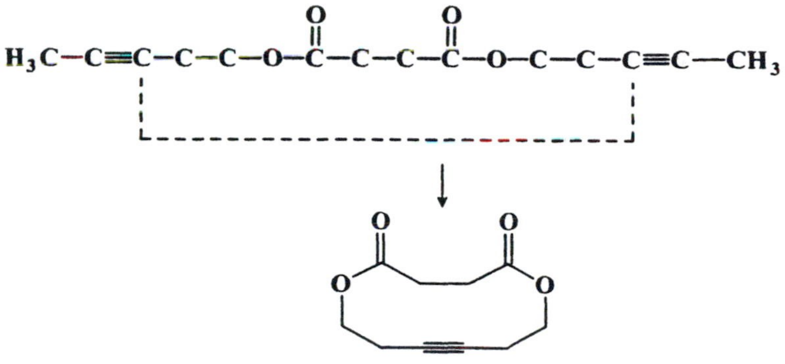

b. $MeC{\equiv}C(CH_2)_8COO(CH_2)_9C{\equiv}CMe$ (with C=O on the ester carbon)

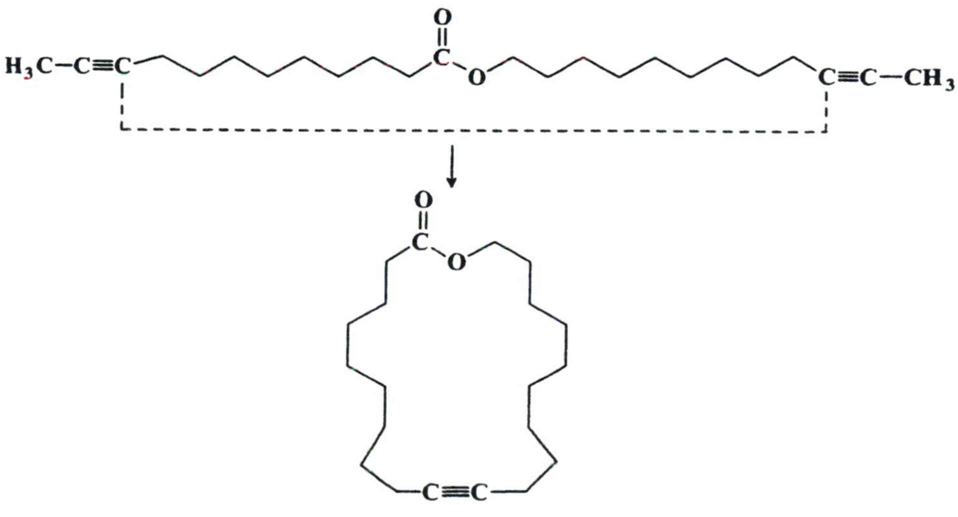

14-26 Evidence in support of the intermediate shown at the right includes:

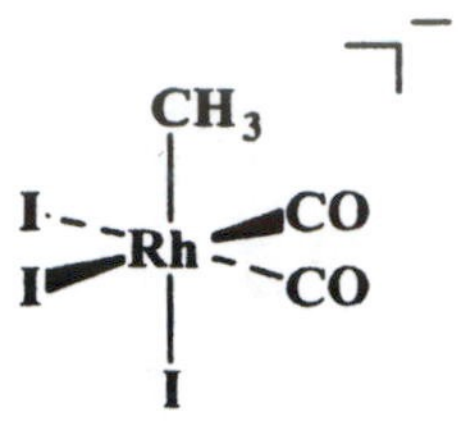

(1) An infrared band at 2104 cm^{-1} has been observed. This band is similar to the higher energy band in the Ir analogue, $[CH_3Ir(CO)_2I_3]^-$, which has carbonyl bands at 2102 and 2049 cm^{-1}. A second band, expected for $[CH_3Ir(CO)_2I_3]^-$, would be hidden under strong bands of the reactants and products of steps 2 and 3 of the mechanism.

(2) The ratio of the absorbance of the 2104 cm^{-1} band to the absorbance of a band at 1985 cm^{-1} of $[Rh(CO)_2I_2]^-$, the reactant in step 2, is proportional to the concentration of CH_3I. This is consistent with what would be expected from the equilibrium constant expression for formation of the intermediate, $K = \frac{[Rh(CO)_2I_2]^-[CH_3I]}{[CH_3Rh(CO)_2I_3]^-}$ and is consistent with the steady state approximation for the mechanism.

(3) The maximum intensity of the 2104 cm^{-1} band occurs when the product of step 3 is formed most rapidly.

(4) When $^{13}CH_3I$ is used, a doublet is observed in the NMR consistent with ^{13}C-^{103}Rh coupling. Other NMR data also support the proposed structure.

Chapter 15
Parallels Between Main Group and Organometallic Chemistry

15-1 **a.** $Mn_2(CO)_{10} + Br_2 \rightarrow 2\ Mn(CO)_5Br$

b. $HCCl_3 + xs\ [Co(CO)_4]^- \longrightarrow$

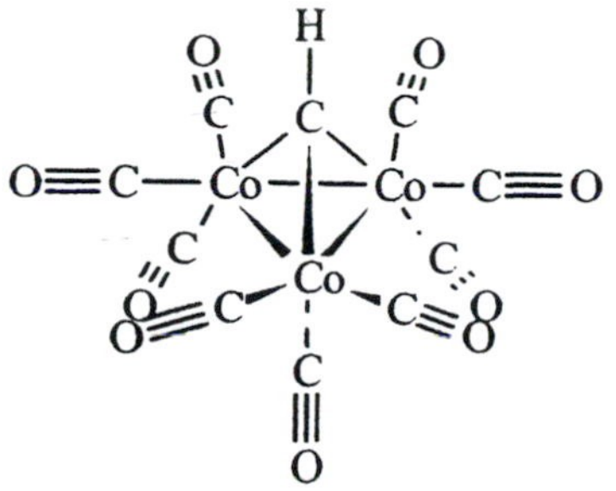

c. $Co_2(CO)_8 + (SCN)_2 \longrightarrow 2\ NCSCo(CO)_4$

d. $Co_2(CO)_8 + C_6H_5\text{-}C{\equiv}C\text{-}C_6H_5 \longrightarrow$

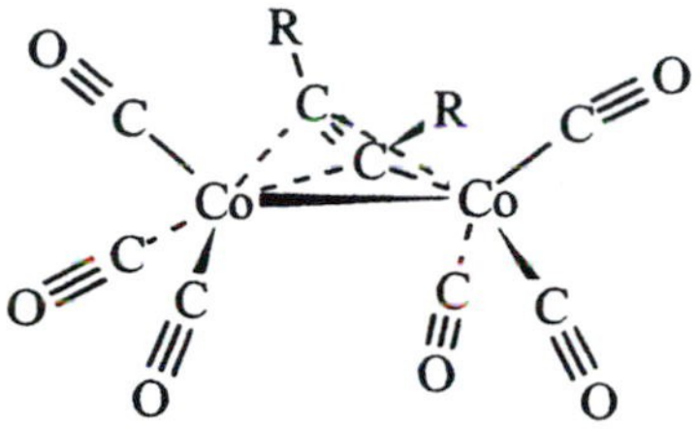

e. $Mn_2(CO)_{10} + [(\eta^5\text{-}C_5H_5)Fe(CO)_2]_2 \longrightarrow 2\ (CO)_5Mn\text{-}Fe(\eta^5\text{-}C_5H_5)(CO)_2$

15-2 **a.** $Tc(CO)_5 \underset{0}{\longleftrightarrow} CH_3$

b. $[Re(CO)_4]^- \underset{0}{\longleftrightarrow} CH_2$

c. $[Co(CN)_5]^{3-} \underset{0}{\longleftrightarrow} CH_3$

d. $[CpFe(C_6H_5)]^+ \underset{0}{\longleftrightarrow} CH_4$

e. $[Mn(CO)_5]^+ \underset{0}{\longleftrightarrow} CH_3^+$

f. $Os_2(CO)_8 \underset{0}{\longleftrightarrow} H_2C{=}CH_2$

15-3 Many examples are possible. Two for each:

a. $CH_3 \underset{0}{\longleftrightarrow} (\eta^6\text{-}C_6H_6)Mn(PPh_3)_2$

$CH_3 \underset{0}{\longleftrightarrow} [Ni(CS)(PMe_3)_2Cl_2]^+$

b. $CH \longleftrightarrow (\eta^4\text{-}C_4H_4)Co(CS)$

$CH \longleftrightarrow (\eta^6\text{-}C_6H_6)Rh$

c. $CH_3^+ \longleftrightarrow$ Mo(borazine)$(PMe_3)(CS)$

$CH_3^+ \longleftrightarrow [V(CO)_3(en)]^-$

d. $CH_3^- \longleftrightarrow (\eta^8\text{-}C_8H_8)Ru(PEt_3)$

$CH_3^- \longleftrightarrow [Co(N_2)(CO)_2(bipy)]^+$

e. $(\eta^5\text{-}C_5H_5)Fe(CO)_2 \longleftrightarrow Co(CO)_3Cl_2$

$(\eta^5\text{-}C_5H_5)Fe(CO)_2 \longleftrightarrow [(\eta^6\text{-}C_6H_6)Cr(CO)_2]^-$

f. $Sn(CH_3)_2 \longleftrightarrow (\eta^5\text{-}C_5H_5)Ir(CO)$

$Sn(CH_3)_2 \longleftrightarrow [Tc(CO)_4]^-$

15-4 Many possibilities exist; some of those given here may not have been made, and may not be stable configurations.

a. $C_2H_4 \longleftrightarrow (CO)_4Fe{=}Fe(CO)_4$

b. $P_4 \longleftrightarrow [Ir(CO)_3]_4$

c. *cyclo*-$C_4H_8 \longleftrightarrow$

$(CO)_4Fe—Ru(CO)_4$
| |
$(CO)_4Ru—Fe(CO)_4$

d. $S_8 \longleftrightarrow$ *cyclo*-$[Fe(CO)_4]_8$

15-5 **a.** All three complexes are 18 electron species, with the benzene ring of $[(C_5Me_5)Fe(C_6H_6)]^+$ replaced by three carbonyl groups or two carbonyls and a phosphine. All have a formal coordination number of six.

b. The experimental results are more complex than might be expected.

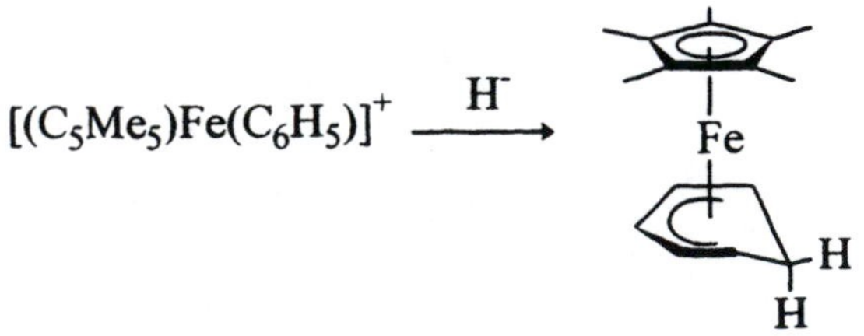

$[(\eta^5\text{-}C_5H_5)Fe(CO)_3]^+ + H^- \longrightarrow (\eta^5\text{-}C_5H_5)Fe(CO)_2H \longrightarrow [(\eta^5\text{-}C_5H_5)Fe(CO)_2]_2$

$[(\eta^5\text{-}C_5H_5)Fe(CO)_2PPh_3]^+ + H^- \longrightarrow$

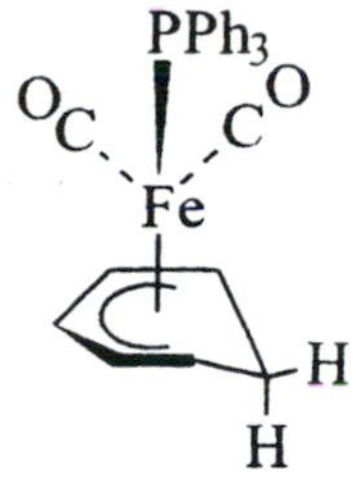

(see A. Davison, M. L. H. Green, and G. Wilkinson, *J. Chem. Soc.*, **1961**, 3172)

15-6 **a.** CH_2, $Fe(CO)_4$, $[Mn(CO)_4]^-$, and PR_2 each has two frontier orbitals, each with one electron. Each fragment is two ligands short of the parent polyhedron (octahedron or tetrahedron).

b. $Fe(CO)_4$ and CpRh(CO) each has two frontier orbitals, each with one electron.

c. $[Re(CO)_4]^-$ has two frontier orbitals, with one electron in each. R_2C is isolobal with two frontier orbitals, each with one electron.

15-7 The $W(Cp)(CO)_2$ fragments are isolobal with CR and CPh, each with three orbitals containing one electron each; PtR_2 and $Cu(C_5Me_5)$ are isolobal, each with two orbitals containing one electron each.

15-8 **a.** If Mn has the lower energy orbital, then the bonding molecular orbital will be made of more of the Mn orbital, so the electrons in the orbital would be polarized toward Mn.

b. The gold orbitals are higher in energy. Rather than matching energies with the lowest of the π orbitals of the Cp ring, they will match better with the higher π orbitals, which have a nodal plane cutting across the ring.

15-9 **a.** CH_2 and $Fe(CO)_4$ are isolobal, each with two orbitals containing one electron apiece.

b. $Mn(CO)_2(C_5Me_5)$ is isolobal with $[Mn(CO)_5]^+$, a 16 electron species, which is in turn isolobal with CH_2. The Mn–Sn–Mo fragment is similar to allene, C=C=C, in which the double bonds force a linear geometry.

15-10 $[C(AuPPh_3)_5]^+$ Structure: slightly distorted trigonal bipyramid, as expected from VSEPR.

$[C(AuPPh_3)_6]^{2+}$ Structure: slightly distorted octahedron, as expected from VSEPR.

In these complexes, each of the $AuPPh_3$ groups can be viewed as having an *sp* hybrid orbital pointing in toward the carbon. Interactions between these hybrids with the *s* and *p* orbitals of carbon give rise to four bonding orbitals in each case. The eight valence electrons available fill these orbitals. The result is the equivalent of four bonds spread over the complex (bond order of 4/5 in $[C(AuPPh_3)_5]^+$ and 4/6 in $[C(AuPPh_3)_6]^{2+}$. Metal-metal bonding is also likely to contribute to the stability of these structures.

For more details, see H. Schmidbauer, et. al., *Angew. Chem. Int. Ed. Engl.*, **1989**, *28*, 463 and *Angew. Chem. Int. Ed. Engl.*, **1988**, *27*, 1544.

15-11 **a.** The $Mn(CO)_5$ fragments have a single electron in their HOMO, largely derived from the d_{z^2} (hybridized with the p_z orbital) of the Mn. Lobes of the HOMOs of two $Mn(CO)_5$ fragments interact in a sigma fashion with the σ_g orbital of C_2 (see Figure 5-5 for the approximate shape) derived primarily from the p_z orbitals of the C atoms:

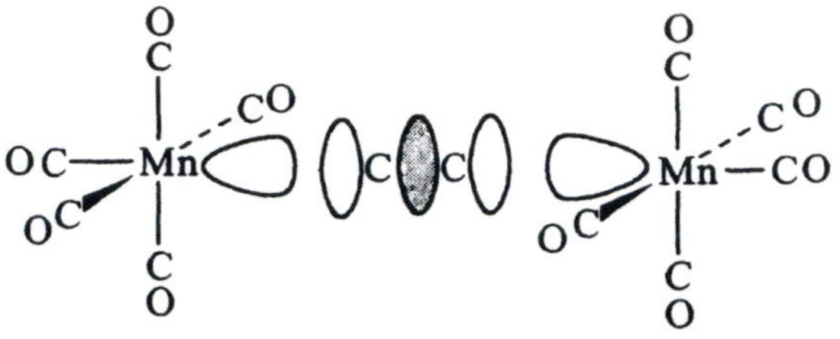

b. The empty π^* orbitals of C_2 can interact with occupied d orbitals of the $Mn(CO)_5$ fragments:

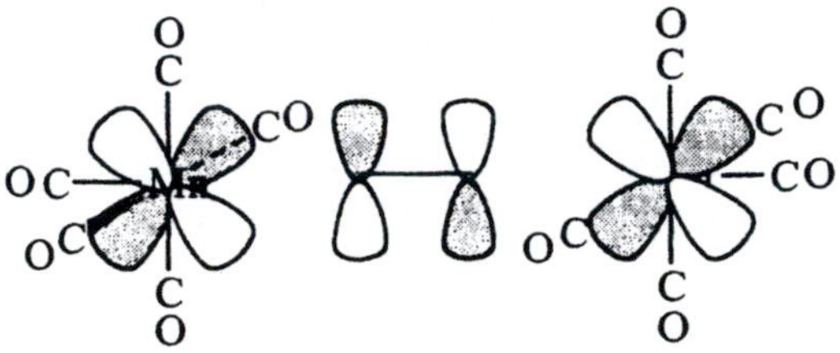

c. The reference points out other interactions, such as between π (bonding) orbitals of C_2 and $Mn(CO)_5$, and discusses the relative energies of the molecular orbitals of this molecule and other molecules having bridging C_2 ligands.

15-12 A staggered configuration is more likely, as predicted by VSEPR. The bonding is similar to that in the triply bonded $[Os_2Cl_8]^{2-}$.

15-13

End view:

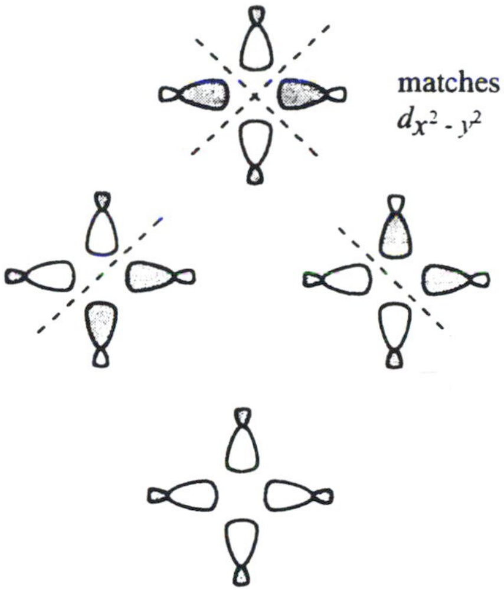

15-14 Cotton's explanation is that removal of an electron effectively changes the oxidation number of Tc, causing the *d* orbitals in $[TcCl_8]^{2-}$ to be smaller than those of $[TcCl_8]^{3-}$. This reduces the orbital overlap, weakening the bonding in spite of the higher formal bond order. The change is small--about 3 pm. See F. A. Cotton and G. Wilkinson, *Advanced Inorganic Chemistry*, 5th Ed., Wiley, 1988, p. 1090.

15-15 As the reference describes, these ions are isostructural and are based on octahedra fused at one mutual face (occupied by three bridging chlorine atoms). The shortening of the Re-Re distance upon reduction is attributed chiefly to two factors: as the complex is reduced, the net positive charge on each metal is reduced (the oxidation state changes from 4 to 3.5), which reduces the metal-metal and with the reduction of the metals, the metal *d* orbitals expand, enabling more effective overlap and stronger bonding.

15-16 This example is similar to the one presented in Figure 15-10. In compound **1** there are six DTolF ligands, each with a charge of 1−. Adding the charges of the bridging hydroxides, the total charge of the ligands in this complex is 8−. Consequently, the total charge on the Mo atoms must be 8+, or a charge of 4+ per Mo_2 unit. Each Mo^{2+} has 4 *d* electrons, giving 8 *d* electrons per Mo_2. Mo_2^{4+}, with its 8 *d* electrons, has an Mo—Mo bond order of 4 (Figure 15-10).

In compound **2** there are again six DTolF ligands. The bridging O^{2-} ligands result in a total charge on the ligands of 10−; consequently the four molybdenums must carry an average charge of 2.5+, or 5+ per Mo_2 unit. There are now just 7 *d* electrons per Mo_2. From Figure 15-10 we can see that Mo_2^{5+}, with 7 *d* electrons, corresponds to a bond order of 3.5. The lower Mo—Mo bond order in **2** results in a longer bond.

15-17 a.

O_h	E	$8C_3$	$6C_2$	$6C_4$	$3C_2$	i	$6S_4$	$8S_6$	$3\sigma_h$	$6\sigma_d$
Γ	6	0	0	2	2	0	0	0	4	2
A_{1g}	1	1	1	1	1	1	1	1	1	1
T_{1u}	3	0	-1	1	-1	-3	-1	0	1	1
E_g	2	-1	0	0	2	2	0	-1	2	0

Γ fits either the *s* or the p_z orbitals.

b. It can be seen from the table above that $\Gamma = A_{1g} + T_{1u} + E_g$. This can also be worked out by the more elaborate methods used in Chapter 4.

c. The reducible representation for the p_x and p_y orbitals and its components are shown in the table below. Figure 15-11 shows the three T_{1u} and the three T_{2g} group orbitals required (which also include some p_z contribution from two of the boron atoms). The T_{1g} and T_{2u} representations are for antibonding orbitals.

O_h	E	$8C_3$	$6C_2$	$6C_4$	$3C_2$	i	$6S_4$	$8S_6$	$3\sigma_h$	$6\sigma_d$	
$\Gamma_{x,y}$	12	0	0	0	-4	0	0	0	0	0	
T_{1g}	3	0	-1	1	-1	3	1	0	-1	-1	(R_x, R_y, R_z)
T_{1u}	3	0	-1	1	-1	-3	-1	0	1	1	(x, y, z)
T_{2g}	3	0	1	-1	-1	3	-1	0	-1	1	
T_{2u}	3	0	1	-1	-1	-3	1	0	1	-1	

15-18 The full symmetry could be worked out in the same manner as on p. 573 for $B_6H_6^{2-}$. The number of orbitals of each type can be obtained more simply by analogy from the results on p. 574:

28 valence atomic orbitals of B combine to form:
- 15 bonding orbitals (2*n*+1) consisting of:
  - 8 framework MOs (*n*+1)
    - 1 bonding orbital from overlap of *sp* orbitals
    - 7 bonding orbitals from overlap of *p* orbitals of B with *sp* hybrid orbitals or *p* orbitals of other B atoms
  - 7 B–H bonding orbitals (*n*)
- 13 nonbonding or antibonding orbitals

15-19 a. $C_2B_3H_7 \longrightarrow B_5H_9 \longrightarrow B_5H_5^{4-}$ *nido*

b. $B_6H_{12} \longrightarrow B_6H_6^{6-}$ *arachno*

c. $B_{11}H_{11}^{2-}$ *closo*

d. $C_3B_5H_7 \longrightarrow B_8H_{10} \longrightarrow B_8H_8^{2-}$ *closo*

e. $CB_{10}H_{13}^{-} \longrightarrow B_{11}H_{14}^{-} \longrightarrow B_{11}H_{11}^{4-}$ *nido*

f. $B_{10}H_{14}^{2-} \longrightarrow B_{10}H_{10}^{6-}$ *arachno*

15-20 a. $SB_{10}H_{10}^{2-} \longrightarrow B_{11}H_{13}^{2-} \longrightarrow B_{11}H_{11}^{4-}$ *nido*

b. $NCB_{10}H_{11} \longrightarrow B_{12}H_{14} \longrightarrow B_{12}H_{12}^{2-}$ *closo*

c. $SiC_2B_4H_{10} \longrightarrow B_7H_{13} \longrightarrow B_7H_7^{6-}$ *arachno*

d. $As_2C_2B_7H_9 \longrightarrow B_{11}H_{15} \longrightarrow B_{11}H_{11}^{4-}$ *nido*

e. $PCB_9H_{11}^{-} \longrightarrow B_{11}H_{14}^{-} \longrightarrow B_{11}H_{11}^{4-}$ *nido*

15-21 a. $B_3H_8(Mn(CO)_3) \longrightarrow B_4H_8 \longrightarrow B_4H_4^{4-}$ *nido*

b. $B_4H_6(CoCp)_2 \longrightarrow B_6H_8 \longrightarrow B_6H_6^{2-}$ *closo*

c. $C_2B_7H_{11}CoCp \longrightarrow B_9H_{13}CoCp \longrightarrow B_{10}H_{14} \longrightarrow B_{10}H_{10}^{4-}$ *nido*

d. $B_5H_{10}FeCp \longrightarrow B_6H_{10} \longrightarrow B_6H_6^{4-}$ *nido*

e. $C_2B_9H_{11}Ru(CO)_3 \longrightarrow B_{11}H_{13}Ru(CO)_3 \longrightarrow B_{12}H_{14} \longrightarrow B_{12}H_{12}^{2-}$ *closo*

15-22 a. Ge_9^{4-} has 40 valence electrons = $4n + 4$. It classification is *nido*.

b. $InBi_3^{2-}$ has 3 + 3(5) + 2 = 20 valence electrons = $4n = 4$. Its classification is also *nido*.

c. Bi_8^{2+} has 8(5) - 2 = 38 valence electrons = $4n + 6$. It is an *arachno* cluster.

15-23 a.
m = 1 (a single polyhedron)
n = 5 (each B atom counts)
o = 0 (no bridging atoms)
p = 2 (2 missing vertices)
8 skeletal electron pairs

b. m = 2 (2 polyhedra)
n = 10 (each B, C and Co atom counts)
o = 1 (1 bridging atom, the cobalt)
p = 2 (2 missing vertices: the top part of the molecule is considered as a pentagonal bipyramid with the top vertex missing, and the bottom as an octahedron with the bottom vertex missing)
15 skeletal electron pairs

c. m = 2 (2 polyhedra)
n = 17 (each B, C, and Fe atom counts)
o = 1 (1 bridging atom, Fe)
p = 1 (1 missing vertices, the top atom of the incomplete pentagonal bipyramid)
21 skeletal electron pairs

15-24 a. The group orbitals are derived primarily from the 3p orbitals of phosphorus and resemble the five group orbitals for C_5H_5 in Figure 13-22. Diagrams of these orbitals can be found in the second reference.

Atomic orbitals on transition metals suitable for interaction (assuming metal centered below P_5 plane):

lowest energy group orbital: s, p_z, d_{z^2}
1-node degenerate pair: $p_x, d_{xz}\ p_y, d_{yz}$
2-node degenerate pair: $d_{xy}, d_{x^2-y^2}$

b. The molecular orbitals of P_5^- are lower in energy than the similar orbitals of $C_5H_5^-$, giving rise to a generally stronger ability of P_5^- to π accept. For example, the energy match between 2-node orbitals and metal d orbitals may be closer in P_5^- complexes than in the case of ferrocene (Figure 13-28), enabling stronger interaction. An example of such an interaction would be between an empty 2-node orbital of P_5^- and a $d_{x^2-y^2}$ orbital of a metal:

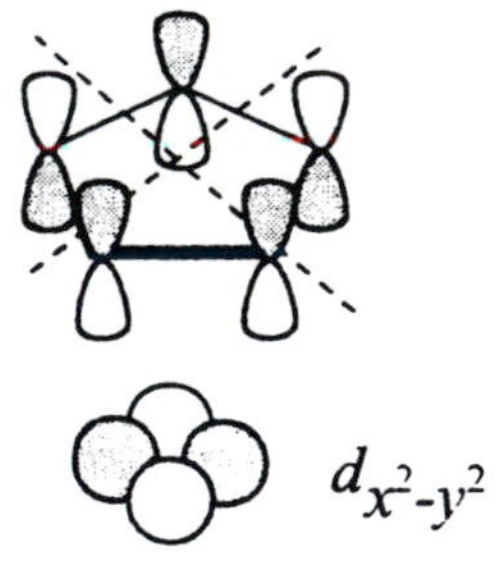

c. The reference provides an energy level diagram of the molecular orbitals of $[(\eta^5\text{-}P_5)_2Ti]^{2-}$. In analyzing the orbitals it is important to see how the π orbitals of the P_5^- ligands match up with d orbitals on Ti and how the resulting shapes illustrate how the lobes of these interacting orbitals merge (in the bonding orbitals) and how nodes are formed between them (in the antibonding orbitals).

15-25 The three types of interaction, σ, π, and δ, should be evident in the orbitals generated. In addition to the bonding interactions (see Figure 15-8) there should be matching antibonding interactions. The relative energies of the molecular orbitals should be similar to those on the right side of Figure 15-9, depending on the level of sophistication of the calculations used.

15-26 **a.** The A_{1g} orbital should have a very large, nearly spherical lobe in the center of the cluster and six smaller, also nearly spherical lobes, on the outside of the borons (centered on the hydrogens).

b. The T_{1u} orbitals should each have two regions where lobes of *p* orbitals on four B atoms merge, plus additional lobes, nearly spherical, centered on opposite hydrogens.

c. The T_{2g} orbitals should show how the lobes of adjacent *p* orbitals merge. The result should have four lobes, somewhat similar in appearance overall to a *d* orbital.

15-27 The orbitals are naturally much more complex, with 167 orbitals and 270 electrons. The T_{1u} and T_{2g} orbitals shown are from a CAChe Hückel calculation.

a. The A_{1g} orbital should be similar to that for $B_6H_6^{2-}$ in problem 15-26, with one large lobe in the center and smaller *d*-orbital lobes on each of the Ru atoms. It is the HOMO, antibonding in symmetry, and the only orbital near this energy that involves carbon orbitals..

b. The T_{1u} orbitals should also be similar to the T_{1u} orbitals of $B_6H_6^{2-}$. There should be two large lobes each derived primarily from *d* orbitals on four Ru atoms and a *p* orbital of the central carbon, plus lobes on opposite Ru atoms (the other Ru atoms). There may also be fragment lobes of *d* orbitals on the Ru atoms. In this calculation, the symmetry of the orbitals is not as easily seen as in the case above.

c. The T_{2g} orbitals are not directly involved in bonding with the carbon, but they do strengthen the cluster framework. The atomic orbitals should interact in sets of four, similarly to the T_{2g} orbitals in $B_6H_6^{2-}$, but some of the distinctive features of *d* orbitals should also be observable in the ruthenium cluster.

Chapter 16
Bioinorganic and Environmental Chemistry

16-1 Methylation reactions can come from methylcobalamin, in which the methyl group is transferred to another substrate. Whether it is as CH_3^-, $\cdot CH_3$, or CH_3^+ is uncertain, but free radicals have been detected in reaction mixtures of this type.

$$LCoCH_3 \longrightarrow LCo(I)^- + CH_3^+$$
$$\downarrow HSR$$
$$LCo(I)^- + H_3CSR + H^+$$

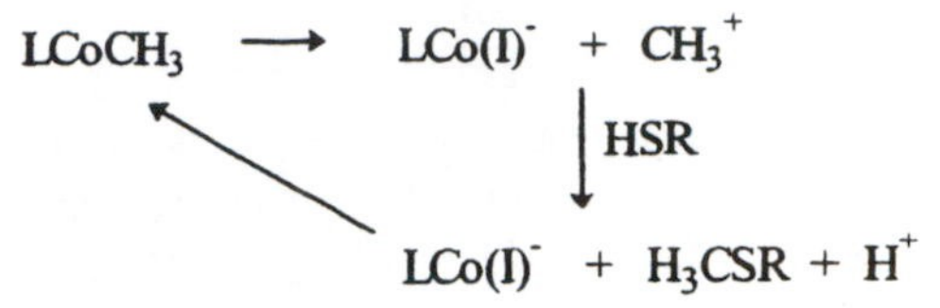

The source of the methyl groups is uncertain; perhaps from methane generated by bacterial action under anaerobic conditions.

Isomerization reactions are thought to require transfer of a hydrogen from the substrate to the 5' carbon of the 5'-deoxyadenosyl group, followed by rearrangement of the substrate and transfer back of the hydrogen. Dehydration of 1,1-diols does not appear to require coenzyme B_{12}.

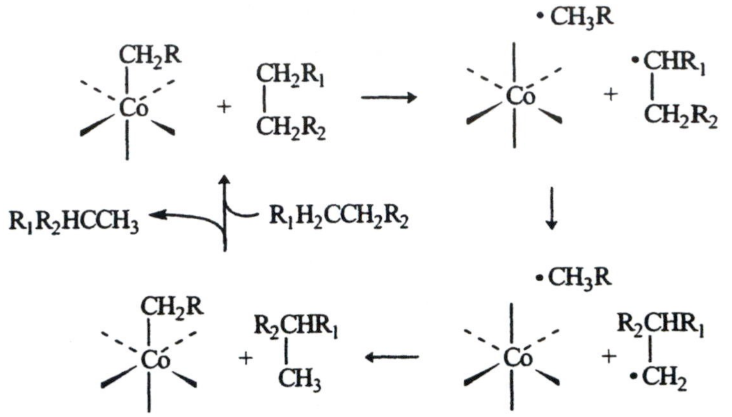

CH$_2$R is the 5'-deoxyadenosyl group; there may be additional R groups on the substrate. Similar reactions can be written assuming ionic intermediates.

16-2 At high oxygen pressure, as in the capillaries of the lungs, hemoglobin picks up a full load of four O_2 molecules because of the cooperative nature of the bonding. At low oxygen pressure, as in the capillaries of the extremities, hemoglobin starts to release O_2, and the cooperative nature of the bonding results in release of all four. Myoglobin serves as a more passive storage device, being less sensitive to changes in O_2 concentration in the midrange. When oxygen is used in reactions, the concentration drops and more is released from the myoglobin. In a resting state, with low demand, it can load up from the higher concentration generated by the hemoglobin release of O_2.

16-3

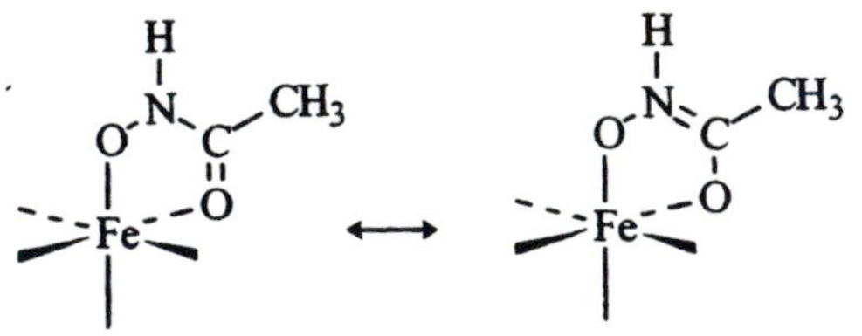

16-4 Hemoglobin contains Fe(II) in the high spin state. When paramagnetic O_2 binds, the result is a diamagnetic complex. Regardless of the description of the bonding, there must be pairing of the electrons of the iron and of O_2 in the resulting molecular orbitals. Many different suggestions have been made, with Fe(II)-O_2 seeming most natural from the starting materials. The structure proposed by Ochiai and shown in Figure 16-4 uses both π

antibonding orbitals of O_2 to form the molecular orbitals. Others have suggested that Fe(III)-O_2^- is more likely, in part because of the oxidizing power of the complex. In molecular orbital terms, all of these are essentially the same, since the result is a set of molecular orbitals, rather than specific components. Mössbauer spectra are influenced by the concentration and symmetry of electrons very close to the iron atom, and indicate the oxidation state of iron. Spectra measured by Tsai, Groves, and Wu show two different electronic environments. They interpret this as being an equilibrium 1:2 mixture of Fe^{2+}-O_2 and Fe^{3+}-O_2^-, with Fe^{2+} increasing at higher temperatures, and the energy levels shown below.

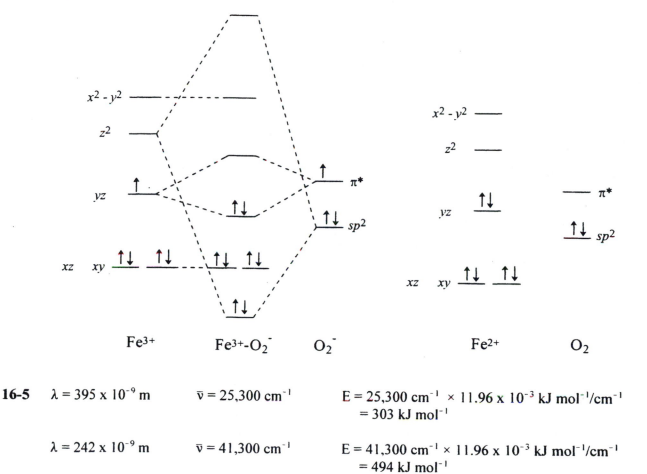

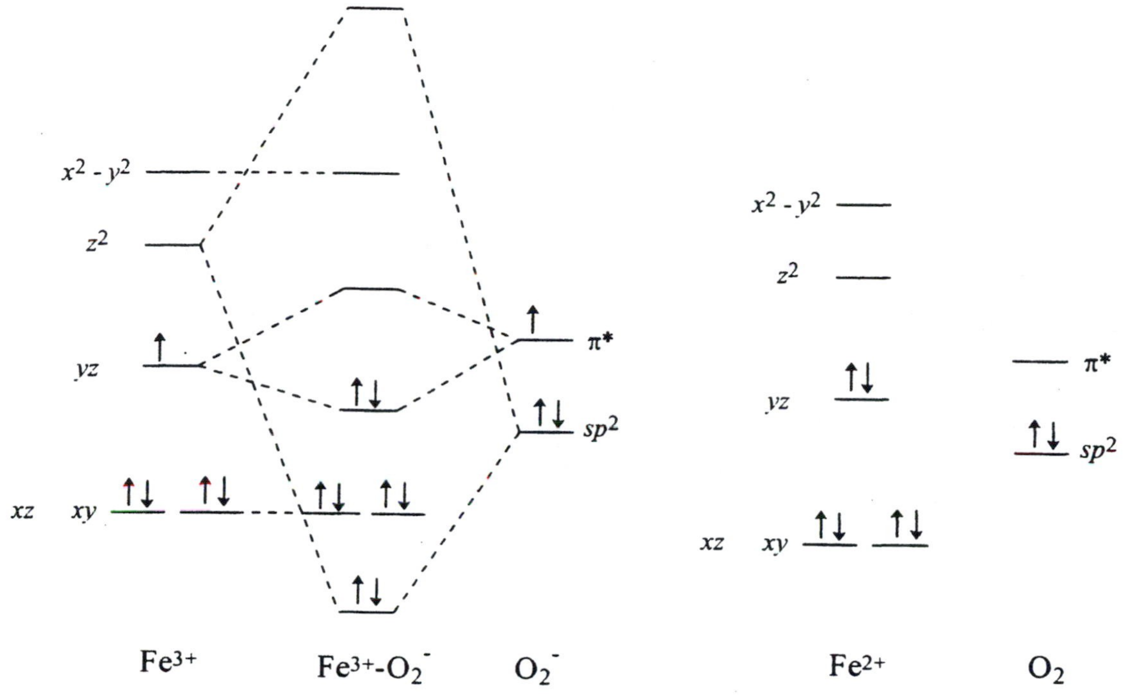

16-5 $\lambda = 395 \times 10^{-9}$ m $\quad \bar{\nu} = 25{,}300\ cm^{-1}$ $\quad E = 25{,}300\ cm^{-1} \times 11.96 \times 10^{-3}\ kJ\ mol^{-1}/cm^{-1}$
$= 303\ kJ\ mol^{-1}$

$\lambda = 242 \times 10^{-9}$ m $\quad \bar{\nu} = 41{,}300\ cm^{-1}$ $\quad E = 41{,}300\ cm^{-1} \times 11.96 \times 10^{-3}\ kJ\ mol^{-1}/cm^{-1}$
$= 494\ kJ\ mol^{-1}$

NO_2 has a bond order near 1.5, O_2 has a bond order of 2. The results here fit the bond order.

16-6 Methyl cobalamin is described as a Co(II) compound, forming Co(I) if CH_3^+ is formed, Co(II) if $\cdot CH_3$ is formed, and Co(III) if CH_3^- is formed. All three have been suggested. As Co(III), methylcobalamin would be a low-spin, nearly octahedral species. After dissociation of CH_3 with any of its three charges, the following electronic structures are possible:

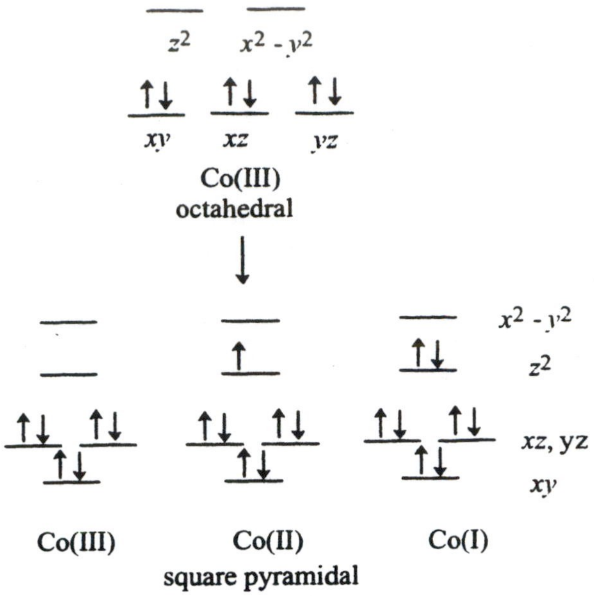

Depending on the separation of the two higher levels, the Co(I) species could have two unpaired electrons, one in each.

16-7 The EDTA-Pb^{2+} complex and similar complexes for other heavy metals are stable enough to work effectively in removing the heavy metals from the body. The danger is in removing useful metals as well (Ca^{2+}, Mg^{2+}, other transition metals). Stability constants for many of these are also high, so all may be removed. Fortunately, the bulk of these other metals is tied up in other compounds and is relatively slow to be released. Since they must be released from other tissues into the blood, their removal is likely to be slow compared with toxic metals ingested only a short time before. If the treatment has been delayed, the problem is more serious because the toxic metal ions are also tied up in tissues and their rate of removal will be similar to that of the essential metal ions.

16-8 If the synthetic blood does not have the ability to absorb excess NO, its use can lead to dilation of blood vessels and low blood pressure, leading to inflammatory disease and degeneration of nerve tissue. On the other hand, if synthesis of NO is decreased because the appropriate enzymes are not present in the blood vessels, some of the same transport compounds are needed to maintain a level adequate to prevent the effects of low NO: hypertension, arteriosclerosis, impotence, and susceptibility to infection.